Axiomatic Theory of Spirituality

For a Happy, Fulfilled and Peaceful Life

My quest for a genuinely spiritual living started as a personal quest though it soon became evident that the quest is universal. I found answers to my quest, and discovered the meaning of my happiness, fulfillment and peace. It is a sweet life that this discovery naturally has enabled for me. There was still work to be done to offer this sweet life universally for everyone. I discovered that everyone is born genuinely spiritual ready for a happy, fulfilling and peaceful life. I traveled around the world until I felt I had smoothed the edges and gained the perspective for universal spiritual living. I was ready to offer a happy, fulfilling and peaceful life to everyone. The presentation of the book was helped by two encounters, one a chance encounter in Sofia and another a sweet echo from the past at Catholic University of Leuven where I had spent two years as a visiting professor. The treatment of spirituality in this book comes entirely from my original research. It answers questions that previously have never been even posed, much less answered.

Axiomatic Theory of Spirituality

For a Happy, Fulfilled and Peaceful Life

Abdur Rahim Choudhary, Ph.D.

MV Publishers

Published by MV Publishers, a subsidiary of Muslim Voice, 12719 Hillmeade Station Dr, Bowie, MD 20720, USA.

MVPublishers@muslimvoice.org

ISBN 978-1-956601-19-0

First edition 2024

United States of America

Choudhary, Abdur Rahim, 1944–
Axiomatic Theory of Spirituality: For a Happy, Fulfilled and Peaceful Life

ISBN 978-1-956601-19-0

To the Seekers of the Spiritual Way

Contents

Preface

My quest for a genuinely spiritual living started as a personal quest though it soon became evident that this quest is universal. I spent five years, from 2010 to 2015, and I did find answers to my quest; I discovered the meaning of my happiness, fulfillment and peace.

It was a sweet life that this discovery naturally enabled for me.

There were still some edges to be smoothed and some perspectives to be generalized in order to offer this sweet life universally for everyone – not just for some but for everyone. I discovered that everyone is born genuinely spiritual ready for a happy, fulfilling and peaceful life.

I took a break for seven years, from 2015 to 2022. I traveled around in Balkans and Turkistan, an area rich in traditions, hospitality, wisdom, and excellent human relations. I absorbed these human values as best as I could. During these travels, I felt I had smoothed the edges in my pursuit of a genuinely spiritual way, and I had gained the perspective for universal spiritual living. I was ready to offer a happy, fulfilling and peaceful life to everybody.

During these years of wandering and seeking I observed and assimilated as well as took notes to myself about my little experiments along the path to discovery. I recently realized that these notes had a value of their own, so I published them under the title "Inner Experiences on the Spiritual Way" with ISBN 978-1-956601-12-1.

However, there was one aspect that still remained, namely that of the style of presentation. That was helped by two encounters, one a chance encounter and another a sweet echo from the past.

The chance encounter took place in Sofia. I had joined a walking tour of Soviet era monuments. There was another tourist in the group who had an acquaintance in the University nearby. I joined him to visit this acquaintance. After a brief tour of the department, we went for dinner where another friend was to join us. His name was Nikolay, a dedicated writer and translator. I shared with him the current version of my book on spirituality, and he provided very helpful feedback.

The second encounter was in Leuven. I had spent two years as a visiting professor at the Catholic University there. Professor Frans Cerulus had now retired but he was still active as Professor Emeritus. I stayed three weeks in Leuven, meeting old friends, especially exchanging ideas with Professor Cerulus on my spirituality book. Professor Cerulus was very helpful.

I still waited another eight years after my visit to Leuven in 2015, before penning down this final version of the book in 2024.

The treatment of the subject in this book comes entirely from my original research. It clearly answers questions that have never been answered previously. The theory is not based on any previous work, so that there has been no occasion to cite references.

Lack of well-defined meaning of the term "spirituality" hinders a scientific study of the subject. Currently there exists no theory of spirituality that elaborates what spirituality is, how a person acquires it, and how its presence within one's own self can be validated.

I have defined a framework for the theory using concepts and terms that are all well-defined, introducing no transcendental elements, and insisting on verifiability. The methodology for this research is derived from the sciences, making the theory scientific, more so than ever before.

In comparison, philosophers like Rene Descartes barely initiated the 'dialogue' between spirituality and science. However, it could not have been scientific because Descartes did not free himself from the influence of the Church: for example, he worked on a proof for the existence of God and immortality of soul. His mind-body duality is expressed under such influences. In comparison, my theory is scientific and does not use religious constructs.

The unique nature of the spiritual entities and events does require an extension of science itself.

I have formulated the foundational concepts, an overarching framework, and a detailed methodology for a formal discipline to conduct scientific research into man's spirituality. This should provide inspiration and impetus to start scientific study of human spirituality as an independent discipline of scientific enquiry in universities around the world.

Please accept this book as a spiritual offering from me to you.

Abdur Rahim Choudhary, Ph.D.
12719 Hillmeade Station Drive, Bowie, Maryland 20720, USA.
RahimChoudhary@gmail.com
January 16, 2024.

Introduction

What is spirituality? Currently, an answer to this simple question does not exist except in relation to religions: and then there are as many answers as there are religions. Thus, an answer exists with respect to what is Christian spirituality, Islamic spirituality, or Buddhist spirituality, etc. In other words, spirituality exists only as an auxiliary subject to religion, but not in its own right as a subject of enquiry, independent of religion.

This essential binding of a discourse on spirituality with religion means that any discourse on spirituality necessarily exists only in the context of that religion, including religious concepts like God and related transcendental elements and ideas.

Even philosophical discussion of spirituality among Western intellectuals necessarily requires God and related religious backdrop. Given the strong influence of Christianity, many modern philosophers in the West have routinely incorporated Christian ideas while discussing spirituality, thereby putting forward a restrictive and limiting philosophical exposure of spirituality with its bindings and indebtedness to Christian religion.

For example, Blaise Pascal in his Pensées, despite his mathematical and scientific distinctions, uncritically incorporates Church's teachings. No surprise that the Church praised Pensées as the best writing in French prose. Indeed, Church's teachings were so impressed on the minds of intellectuals that concepts like God and soul are taken to be valid without critical examination and evaluation. Thus, Henry Bergson makes unapologetic use of them while presenting his thesis on Two Sources for Morality and Religion. He does not hesitate to suggest a Christian society as a valid source for morality. This has led to acceptance of a rather restrictive and limiting understanding and knowledge of philosophical discourses on spirituality, namely those restricted to Christianity. Marguerite Porete's work "The Mirror of Simple Souls" is portrayed as a work in mysticism though it presents hardly anything other than Christian ideas. Eckhart is presented as a Western philosopher, while in fact he speaks as a devout Christian clergy. Many people are celebrated and put forward as symbols of Western ideas and free and scientific thinking, whereas they simply mirrored what Christian ideas about spirituality were at their times.

I have removed such limitations in my treatment of spirituality. To start with I treat spirituality as an independent discipline of enquiry. That means to decouple spirituality from religion on the one hand and religious artifacts on the other. An example of religious artifact is the concept of God and soul or equivalently some such transcendental entities. While I do not need such concepts and constructs in my treatment of the discipline of spirituality, the resulting academic output is for the benefit of all – religious or not, godly or not. The reason is that the outcome of my scientific enquiry is independent of such aspects, so that it will show a way to happy, fulfilled and peaceful living for everybody.

What I offer in this treatise is spirituality as a scientific discipline of enquiry like other disciplines such as history, geography and biology, etc. When I say scientific, I mean applying scientific research methodology to spiritual research, though I will show that the current scientific tool box is rather limiting and does need to be expanded in order to research entities and events that populate the spiritual world. Such expansion is also necessary to enrich life in the physical world by saving mankind from its own foolish greed and making his life happy, fulfilling and peaceful, not only at individual level but also at collective levels.

Doing so frees a person from the disadvantages of ambiguous and opinion-based thinking because scientific thinking is free from those. A person almost immediately begins to reap the advantages of this approach in terms of a happy, fulfilling and peaceful life that evolves and grows instead of a static kind of happiness that is often marred with ambiguities and shifting ground.

The entirety of existing literature on spirituality is squarely indebted to religion. It is full of religious artifacts such as soul, spirit, angels, devil, saints, and divine, etc. If I want to decouple the discourse on spirituality from religious concepts and artifacts, it would require that the discipline of spirituality be developed from scratch – starting from its definition, its attributes, its inculcation, its verification and validation, its overarching facts, journey of its discovery, fruits of its presence, proofs of its validity, ways to research it, the long infinite pursuit into oblivion, fruits of wisdom, ecstasy that fills life, and constant wonderous surprises along the way.

The theory of spirituality that I present is aimed at necessarily to begin to make life happy, fulfilling, and peaceful. If it does not do that for you, that is the time for you to pause and reevaluate. It is because a journey that misses out on happiness, fulfillment and peace is not spiritual. As presented in this treatise, spirituality is different from anything that has so far been written under the topic. In particular, it is not religious, it is not something divine, transcendental or mystical. Rather, it is totally immanent and scientific. As such things like belief or faith are not needed, though people with belief and faith equally benefit from it.

It is not what the philosophers have written under this topic, because, until lately, such philosophers have been under the influence of the Church. It is also not the thing that gets used under the name spirituality, rather loosely and uncritically, in areas such varied as medicine and architecture.

Spirituality has nothing to do with spirits. It has nothing to do with souls, angels, ghosts, daemons, paradise and hell. As I stated earlier, this topic is entirely immanent and scientific.

Spirituality starts when you look inwards, you talk to your own self and get to know your own self. It is reflective. You ask your own self what happiness means to you, and how to be happy? And it tells you the answer to your enquiry. Likewise, you can converse with your own self about fulfillment and peace; and it will tell you the answer to your enquiry. You can converse with your own self anything that occurs to you, and you will receive the answer you seek.

It is important to note that you alone know your own happiness, no one else is privy to that, and no one else can show you how to achieve it. Therefore, you keep your happiness in your own hands so that you can know it and achieve it, and also find out and correct humanly errors and omissions in the process. Those who seek happiness in the outside world, they often struggle without reaching because they lack inner understanding and transparency of the process.

The answers you seek often do not come instantaneously or clearly. There are two reasons for that. The obvious reason is the camouflage that the society places on your own self so that the answers, that might appear to come from your own self, are really reflections of the artifacts of the society. Your self warns you against living an unexamined life, therefore it is an important task for any spiritual seeker to look inward and examine his own state and remove this camouflage, so that it becomes easier to consult with his own self. The second reason is the state of wisdom relative to what is required to answer such query. This state of wisdom is a dynamically moving state that improves with the extent of your spiritual advancement. I will say more about it in due course, in the chapter titled Acquiring Wisdom on page 57.

Getting to your own self is not simple or easy. That is because the self is camouflaged by artifacts of the society in such a way that while you think you are consulting your own self you may be consulting this camouflage which represents the artifacts of the society rather than your own self. Some effort is needed to be able to reach your own self. There are procedures and processes to use for this purpose, which I will elaborate in due course, in the chapter titled 'Spiritual Journey: An

Illustration' on page 83. Let it suffice for now that it is important to guard against the artifacts of the society, and to experiment with them while exercising self-reflection.

From this point of view, your own self is your guide. To heed this guide is totally a spiritual thing to do and is essential for the spiritual living. Spiritual living is also referred to as the spiritual journey, because it is a process, not a destination; the process is infinite and nobody reaches its end. Everybody who seeks a spiritual way is a traveler along this process.

Your own self is the only guide who can lead you in spiritual ways. Short cuts like asking someone else to lead and guide you will only complicate the pursuit and possibly subject you to exploitation and even outright fraud. So, trust your own self and use due guardedness and persistent and consistent effort at conversing with your own self. The journey is zigzag because you will miss the points while consulting with your own self, and even make outright errors. However, that is part of the journey. The process makes sure that you will naturally get an opportunity to correct your error; and the process gradually prepares you and strengthens you for a spiritual living that is sure to bring happiness, fulfillment and peace. For this process to work, it is crucially critical that you use your own self as your guide versus delegating this function to someone else. A reassurance that you are on the spiritual journey is the happiness, fulfillment and peace that it will bring to you, not eventually but right away, though its quality and nature will evolve throughout the journey. I will further elaborate in due course, in the chapter on Prerequisites for Spiritual Beginners on page 69.

Spirituality is the inward-looking view, also referred to as the spiritual world. Things in the spiritual world consist of inner states of man such as the feelings of happiness and wretchedness, peace and anguish, fulfillment and desperation, passion and rancor, pleasure and pain, ecstasy and depression, and so on. Unlike the physical things, they cannot be formulated in terms of the space-time constructs that science uses. These constructs are used for things like weight, distance, area, volume and speed etc. However, the spiritual entities mentioned above are of an entirely different type.

Man needs to discover the spiritual world because his happiness, fulfilment and peace depend on it. Therefore, man has an innate urge to discover the spiritual world. He has a natural tendency to look inward, though the constraints of the society persistently and consistently keep him occupied in the external directions.

Religious spirituality is one of those external directions. It is readily recognized as such because of explicit presence of God, soul and spirits, etc. In contradistinction, many existing and quite prevailing exposures to spirituality are esoteric – they generally originate from Hinduism. They appear to be non-religious and at the same time retain the artifacts of the religion including its mysticism and numina such as Super Beings and Souls. This kind of spirituality is a camouflaged variety of religious spirituality. It is hard to recognize it as such, given its claim to be non-religious, and given the fact that its proponents go to great lengths to obscure and hide its religious nature. Such approaches tend to reduce spirituality to a sort of personal therapy. It violates the fact that spirituality is inward looking view and it is not externally induced.

It also violates the above-mentioned principle that your guide is your own self, and it is unsafe to delegate this role to someone external, such as a guru. Moreover, it treats spirituality as something personal, and avoids collective spirituality towards world harmony and wellbeing, as well as confronting the destructive side of mankind.

There is another aspect of spirituality that is, on the surface, convincing in its appeal. It convinces the seeker that spirituality is about meaning, such as the meaning and purpose of life, and the quest for a fully integrated approach to life. This type of quest for meaning is divertive for multiple reasons. First is the fact that after having raised the questions, apparent attempts to answer them are incomplete and inconsistent. Such attempts use terminology that is ill-defined and using this confusion they introduce transcendental factors to invoke some sort of mystic numen. A serious analysis will reveal that the questions that are raised are the same or similar to the ones that religions had already raised in order to plant them in our thinking process. The answers provided are ineffectual.

There is a recent trend that is popularly called "spiritual but not religious". It seems to have arisen by a general distancing of the people in the West from their religion. At the same time, they did not want to distance themselves from the spiritual aspect of life. Therefore, they were attracted by a notion where they could free themselves from religious dogmas and yet get the benefits of a spiritual life.

This trend is not systematic, and there is no clear idea about what spirituality is. While people dropped the dogmatic and ritualistic aspects of their religion, most continued with the spiritual aspect. The spirituality they continued to use is therefore religious spirituality. The trend is quite diverse with respect to what population is attracted towards it, ambiguities about their ideas of spirituality, and the role that such an individualized religiosity plays in life. Moreover, it lacks interpersonal relations within a community, and is vague about happiness and peace.

The discipline of spirituality in this book is axiomatic as is explained in the chapter on Axioms Used in This Treatise on page 29. All meanings and processes naturally evolve out of living life itself, with one's own self as the guide. It is entirely well defined and scientific; there is no ambiguity, uncertainty and mysteriousness. However, the current toolset used by science is inadequate for research in spirituality, and it does need to be extended minimally by two additional elements, namely internalization and witnessing. My approach, therefore, contrasts with the existing approaches to spirituality which are ad hoc and eclectic.

Spirituality, or the spiritual world, is the inward-looking view; and physicality, or the physical world, is the outward-looking view. Man needs to discover the physical world because his life depends on the resources within it, like air, water, food, and shelter. Therefore, man has an innate urge to discover the physical world. Since the inward-view and the outward-view are together exhaustive, there is nothing else in the observable universe other than the spiritual world and the physical world.

A child's curiosity with everything around him illustrates this. There are three levels of the discovery of the physical world. They respectively satisfy man's biological, social, and intellectual needs. The innate senses and intellect of man enable these discoveries. Man has used the innate senses and intellect to develop the methodology of scientific research for a deeper discovery of the physical world.

Does a spiritual world really exist? Does a physical world really exist? Both worlds exist as long as an inward-looking view of the universe exists, and as long as an outward-looking view of the universe exists. The definition of both worlds is symmetric with respect to each other. One exists as assuredly as the other.

Are the spiritual world and the physical world totally different? As I will demonstrate in due course there is an equivalence principle that operates between the two worlds such that the entities in the spiritual world can be translated into entities in the physical world, and vice versa.

Spiritual world implies man to consult his own self, for example to know what his happiness means and how he can acquire it. Similarly, he gets to know the meaning of fulfillment and peace and how he can acquire these. Confusion can arise for some people because there are apparently similar things in the physical world that deceptively resemble happiness, fulfilment and peace. For example, there are the concepts like pleasure, security, and success. Physical world can provide these in ample. However, pleasure is not happiness, security is not peace, and success is not fulfilment. Therefore, a person can be pleasured and yet unhappy, he can have excellent security and yet enjoy no peace, and he can successfully achieve the most prestigious position and yet remain unfulfilled. Only the spiritual world can guide and satisfy a man in his pursuit of happiness, fulfillment and peace. No matter how successful man becomes in his discovery of the physical world, he remains unfulfilled until he also discovers his spiritual world. Spirituality brings man happiness, fulfillment and peace as well as pristine values and immanence to his life with a connectedness amongst all.

For these reasons man has a natural urge to discover both worlds: he seeks to discover the physical world because his biological survival depends upon it, and he seeks to discover the spiritual world because his happiness and peace depend upon it. Both worlds exist in their own right without one of them dominating.

Survival and Happiness

A person is dependent on his society for many things that he needs for his survival. This dependence arises from the fact that the society controls the resources that a person needs for his survival. Certain special interests in the society often enjoy much of this control. I will sometimes use the term society, but I will mean the special interest groups within the society that enjoy much of the control of its resources.

The dependence on the society starts early on in life and it lasts throughout life. Right from the birth, a person is embedded in the society, without even knowing what society is. His first exposure to the society is through his parents, his siblings, the friends of the family, their siblings, and others with whom his family socializes. His exposure to the society is also through the clothes his parents make him wear, the foods his parents feed him, the bed he sleeps in, and the house and community that his parents live in.

The parents and relatives teach the child. This teaching is often full of love and it is well intentioned. However, it also includes aspects regarding the impact of greed and corruption. That can hinder the child's own self by subjecting the child to conformity with the values of the society. This hurts the child in two significant ways: first, it plants values in the child that may not be conducive to further growth of his own self; second, it makes the child conformist versus the free spirit that the child was at birth.

This challenge only gets more formidable, and ever more entangling, as a person interacts with the society more and more. He is sent to kindergarten, and then to school. Next, come the long years of a career where there are added factors like competition, intimidation, and the long arms of the big brother. He gradually acquires the ways of the society, with little opportunity to take a pause and reflect, to think and discriminate, and to look inside his own self to feel the touch of his innate spirituality.

Man has four fundamental needs. The first category consists of those things that are necessitated by his biological survival: food, clothing, and shelter. Second is the need for sex that is necessitated by the law for continuation of species. The third category consists of those things that are necessitated by his social needs in the society in which he lives, for example, seeking a prominent position in the society and a desire for fame and fortune. The fourth category is necessitated by his innate spirituality: seeking an inner meaning to his happiness, fulfillment and peace. The items in the first category are essential for pursuing those in the second and fourth categories, while items in the third category are generally a hindrance. The society controls the resources for the first three categories, and the society distracts a person enough that he overlooks his needs in the fourth category. When man overlooks his spiritual needs, he deprives himself of happiness, fulfillment and peace.

On the minimum, every person needs to meet his subsistence needs. It is his choice regarding how much resources a person wishes to acquire for his biological subsistence. More of the resources he wishes to have, more is his dependence on society, and more influence society will exercise over him. A person must perform this tradeoff diligently.

Some people choose minimal resources for their subsistence and maximally preserve their freedom from the influence of the society. The society usually casts such people to its fringes and portrays them as failures. It is no wonder that very few people make this choice. Examples of such people are some Sufis and some Saints as well as some poor people who quietly enjoy their freedom away from the limelight of the society.

On the other hand, some people choose maximal resources for their subsistence, which means minimal freedom from the influence of the society. Most people make this choice because of greed. A vast majority of such people spend their lives under duress because, no matter how much they struggle, their resources remain small and their freedom gradually vanishes. A vast majority of people in a society fall under this class. For them both the resources and the freedom are mirages.

A few people choose maximal resources and actually manage to attain them. The society regards such people as successful and gladly admits them into a privileged class. They often constitute upper few percent of a society, and usually belong to the special interest groups. By choosing maximal resources, such people have chosen minimal freedom from the influence of the society, and that is what they get.

Even though they are members of a privileged club that can influence the society, they still have only a minimal freedom from the forces of the society. The society has its own strings to tie the hands of those whom it admits as members of its privileged club. Any individual member probably weighs less than one percent of the club. Despite being privileged in the society, the members of the club got themselves admitted due to their special qualities such as extreme competitiveness, ruthlessness, and suppression of the voices of their own self. Therefore, these people get what they bargained for: they enter the privileged club of the society, but they do not regain their freedom from the society. This holds for the rich and powerful members, including the kings and the high priests.

Wisdom is necessary to decide what a person wants to choose with respect to the resources. Wisdom is a spiritual thing. A wise man will consult his physical world and his spiritual world and keep his decision free from greed. Everyone makes the tradeoff according to the state of his wisdom at the moment of the decision. Everyone chooses to aim for certain resources, and consequently pays for it by accepting a proportional dependence on the influence from the society.

No matter what choices a person makes, he will end up not doing three types of things in his life. The first category comprises things that he was not even aware that he could do. How does a person remain unaware of these things? That is because the society evaluates his social position and tells him who he is. In other words, the society assigns him a role and expects him to play it. This is equivalent to the society placing him in a restrictive box, within which he lives his life. The role in which he is placed imposes limitations on him. These limitations restrict the scope within which he can exercise his senses and his intellect. A person allows this to happen when he is more concerned with his physical and social needs compared to the needs of his own self. The needs of his own self are his spiritual needs.

The second category consists of things that he was aware of as possibilities, but he could not do them because the role that the society assigned to him did not enable him to do them. He does not enjoy the privileges that the society reserves for a more prestigious role. Such privileges are glorified within the society in order to keep a person tempted to spend his life in achieving just a few roles of progressively increasing privileges. He could not do the things that he wanted to do because he did not achieve a designated role in the society that had the appropriate privileges.

The third category comprises the things that he was aware of as possibilities, and the society had empowered him to do them, but he chose not to do them because they were not in line with his desire to achieve the next higher privileged role in society. For example, he could treat his loved ones better but he passed the opportunity in favor of his struggle to move into the next higher privileged role.

Only attention to the spiritual needs will liberate a person from such limitations and errors. However, society keeps a person so engrossed within itself that he neglects to take a pause from society dictated activities and attend to his spiritual needs to find out what his own self desires. This is a tragedy of life that inflicts most injury to a man's happiness.

Its antidote is also clear. A person should stop being greedy; he should not be overly dependent on the forces of the society; and he should remain connected with his own self, in all that he does or does not do.

The antidote at the community level is to have spiritual communities which is the way to global peace and prosperity, as I will discuss in the chapter titled 'World Peace and Prosperity: Final Milestone' on page 87.

Corruption in Society

All life has an innate force that seeks its own preservation and growth as a specie. The force is intended to help life towards its fulfilment.

Benevolent Force: There is only one force that is innate to man, namely the inner urge to self-preserve and grow. Like all the innate things that man has, this force too is a benevolent thing.

The force of self-preservation shows a new born how to eat and exercise his body, senses and intellect. It shows everyone to seek food and shelter. It tells a youth to seek a sex partner. These are all benevolent manifestations of the force. As man experiences such manifestations, he also can experience happiness, peace, and fulfilment. The life of a man is a balancing act between his nature, his innate capabilities, and this benevolent force.

Greed Force: Man can use this benevolent force non-optimally, giving rise to other forces that are destructive for his self-preservation as a specie and his happiness, fulfilment and peace as an individual.

The benevolent force remains benevolent as long as man uses it naturally, in accord with his conscience, his own self. When man uses the force in discord with his own self, the force is benevolent no more. Therefore, man must always listen to his own self in living his life.

Man uses self-preservation force unconscientiously when he is not content with the benevolent experiences of this force. Rather, he wants to maximize this experience versus using it at its naturally optimized level. Therefore, he wants more and more food, shelter and sex. The more he wants these things the greedier he becomes. Greed happens when a person wants these things at the expense of others.

Greed is an errant manifestation of the benevolent force. It is so because man does not use the innate force optimally; rather he seeks to use it maximally and, in the process, seeks to possess what is the share of others.

Greed becomes a disease in man, and it is his own doing. It is primarily a mental disease, representing an imbalance of thought that corrupts his worldview. For example, he usurps what is the share of others. Greed is to a man as termite is to wood. It will eat into his systems of happiness, fulfilment and peace. Therefore, a man must always ask if he is being greedy. His own self should guide him, versus him being led by the trends in society.

Epidemic Greed Force: The greed force multiplies when more and more people in society suffer from it. The aberration that was a disease in a person now multiplies and spreads to many others. It becomes an epidemic in society when majority of the people become infected. It reflects a lack of basic values in most people in society due to a derangement of the collective thought.

When greed becomes widespread it becomes acceptable under the norms of society. As a result, individuals in society fall deeper into greed. The epidemic greed force represents the state when greed is widespread within a society because it becomes acceptable under the norms of the society.

A man should not be misled into accepting greed just because society has accepted it. The safeguard is for man to consult his own self. Consulting someone else in the society can be unsafe. You save yourself from greed because you do not want to destroy your happiness, fulfillment and peace.

Corruption: When the society accepts greed as legitimate because of the epidemic of greed, it opens the doors to corruption. Certain special interests develop within the society who seek to benefit from this corruption, and they take ownership of its management. They know the nature of corruption, but they seek to benefit from it against all advice from their own selves.

Special interests turn their survival needs into a love for possessions. In the process they begin to love possessions versus loving people; to satisfy their greed, they exploit unsuspecting people in the society. The epidemic greed force thus becomes the corruptive force in the society.

Special interests will go to great lengths in order to convince everybody that their corrupt practices are legal. In the process, they corrupt the pillars of society like the judiciary and the church. They portray it as the trait of the successful and powerful. They will decorate it as a mark of distinction. However, in reality it breeds selfishness, hypocrisy, exploitation, and deceit.

The special interests are a small minority in a society. However, they have a large number of followers and employees. A vast majority remains out of the special interest groups. This majority falls prey to the exploitative practices of the special interests.

The greedy and corrupt special interest groups act in discord with their own selves and therefore they cannot enjoy happiness, fulfilment and peace, despite their tremendous possessions and the large influence that they enjoy in the society.

Dehumanization: A society itself is like a box in which its members live. Another society is also like a box in which the members of that society live. Each of these boxes have limitations on them, like the total wealth that the society controls, and the total land and resources that the society can use. A society takes its wealth and resources and partitions them among the different roles that the society has decided to let flourish. Different societies have a different take on how it partitions its wealth and resources, and what roles it allows to thrive. By design, some roles within a society are more privileged. Society glorifies these privileged roles. These privileges, however, do not usually promote a person's happiness, fulfillment and peace.

Greed and selfishness lead the special interests to exploit the silent majority using corrupt methods. Each society has its own brand of corruption, and it uses this brand to expand the boundaries of its corrupt practices to exploit other societies. That is because in their own society there are not enough people left for exploitation, and greed demands more. This leads to conflict among societies, with each society pursuing to exploit in order to satisfy their greed, and in the process to expand the depth and scope of their corrupt practices.

In the midst of this corruption, the warmongers thrive on inter society conflicts. They create conflicts where none presently existed. They develop lethal war machinery, claim to be on the side of the righteousness, and demonize their opponents. In a world where mighty can claim to be right, the warmongers take the license in their hand to paint a deceptive self-portrait and to dehumanize their opponents. They dehumanize whole populations in an effort to make their wickedness look like a virtue.

Dehumanization has given rise to racism and bigotry.

Worldview and Values

Our worldview is the way we view the world; what is in the world, what happens in the world, and our attitude about these.

The worldview of a person is based on his values, knowledge and its internalization, and thus his spiritual state. As these factors change over time, so his worldview evolves. Cumulative effects, up to any point in his life, form the worldview of a person. People do things and expect results according to their worldview.

A small change in the values of a person generally produces a large effect in his life. The effect is positive if the change is positive. For example, his life becomes happy and peaceful, and things that were a source of stress and complex problems tend to resolve themselves into relaxation and calm. If the change is negative, the person will have his happiness and peace reduced. His stress level will increase and the problems in his life will become more bothersome. The processes can take time but the person will begin to feel the effects almost immediately. This is how critically important are the values and worldview of a person. His happiness, peace, stress level, problems, and pressures in his life are the barometer of his wellbeing. When he feels unwell despite his continued struggles that is a warning that something in his worldview is not right. When a person feels happy over prolonged times that is a confirmation that his worldview is pretty healthy and it is also an invitation for enhancement in order to make his happiness genuinely dynamic. A genuinely dynamic nature of his happiness is an important litmus test of his worldview being largely informed.

It is important to realize that while the senses and the intellect are innate to a person his worldview is not because it is the result of exercising the innate capabilities which involves interaction with his environment. One critical environmental factor is the influence of the society on the individual. For example, the society can influence a person to think that he must submit to authority, that there is an assigned purpose to his life, that transcendental elements exist in his life, that he must not question the veracity of certain truths, that he must not indulge in pleasures, that his worldly life is merely an illusion, and other similar ideas. The society may indoctrinate a person from the early childhood so that such ideas become part of his worldview, without any critical appraisal on his part. As he grows wiser, he may let go some of these ideas. This can happen for multiple reasons; some ideas change because the society itself has changed and does no longer hold those ideas as true, scientific progress may force to change certain ideas in a society, and a person may let go of an idea when it conflicts with his experience or his conscience.

If a person adopts a worldview uncritically, usually under the influences of the society, he can have a wrong understanding of what happens in his life, why it happens, and how to respond to it. For instance, a person may think that a particular idol (for example money) that his community worships (highly values) will make things happen in his life. When such a person comes across a misfortune in his life, he may not try to analyze why the misfortune happened; instead, he may attribute it to the idol, and redouble his worships to please the idol. Such blind acceptance can be a prescription for corruption, exploitation, and misery. For the life conditions of a person to change, the worldview of the person must change, which includes a change in things that he regards as true. That is because every person thinks and acts according to his worldview. He uses his worldview to make sense of life, of events that happen around him, and to plan his future actions.

It is a useful spiritual technique to focus on the state of innate capabilities and how well a person uses them. I will refer to this technique as "recall of the innateness". Doing so can often point out the pathways through which unnatural and corrupt things can enter the worldview. For example, the genuineness of the use of innate capabilities may be affected by the society mores, the difficulties of everyday life and techniques deployed by the special interest groups like politicians and clergy.

Recalling the innateness helps clear the mind and free the self from the dogmas of the society. It helps a person determine what part of his worldview comes from his own self and what part is extraneous. He can then take active steps to cleanse his worldview of the extraneous, and reinforce what is from his own self.

The desires, actions, and struggles of a person result from his worldview and they serve as a driving force for his life.

Pristine desires and determined courageous actions to achieve them uplift a person with unrestricted potential for high ideals. It is this aspect that makes some individuals to serve mankind in distinguished ways. Talent and intelligence are fairly evenly distributed, but wise passions and ecstatic efforts make all the difference. The difference is visible in physical aspects and is due to the spiritual excellence.

It works in the negative direction as well. Greedy and corrupt desires and efforts that violate human values can sink a person to the depths of bottomless abyss.

A persistent and intense desire can keep a person restless. A persistent and intense action can keep him busy. The worldview of a person speaks through his desires, actions and struggles. If desires are keeping him constantly restless and unhappy, and his actions are keeping him too busy to take a pause and consult his own conscience, then it indicates the need to cleanse his worldview.

Axioms Used in This Treatise

I discuss here the axioms that this treatise uses, along with the context how each is used, why it is needed, and the role it plays in the theory of spirituality. Further comments on these axioms are made in the chapter that follows. Reference is made to each axiom when the axiom is invoked in the theory, as it is discussed in the subsequent chapters.

Axiom 1

Statement:

Man exists.

Man possesses the following two innate capabilities that help explore the physical world: senses of touch, taste, smell, hearing, and sight; and intellect, including intelligence and mind.

The senses are like sensors for making observations. Intellect analyzes the observations, and helps hypothesize and theorize the results of the analysis.

A man directly observes the physical world through his senses and interprets the sensual perceptions through his intellect. Senses detect the presence of various attributes of the physical things. The eye detects the visual imagery of things, ears detect their acoustic aspects, the tongue detects their taste, the nose detects their odor, and the hands and feet detect the feelings of touch. There are physiological and mental processes that lie hidden behind the senses. For example, the eyes see

29

the objects, but the process of seeing is impossible without the circuitry between the eyes and the brain, and the inherent logic that processes the optical impressions to produce the result of visual observation. The hearing is similarly impossible without the communications between the external ear and the brain. Similar is the case for other senses. For the senses to work, it is required that some relevant biological features be present and that the intellect must correctly interpret the sense perceptions. The sensual observations are not complete without the intellect.

The intellect processes much more than the sensed signal. It can perform correlations, analysis, and theorization. It can process the imagination. It can imagine unseen things and conduct an abstract analysis, formulate intricate set of equations that describe a theory, and contemplate possibilities that have not been experienced before.

Man is predisposed to make **observations.** Even a newborn observes his surroundings. Man observes what is happening and 'measure' its effects as part of the observation process. The purpose of making observations is to discover man's environment, usually through experiments. Examples of the physical entities to be observed include the following situations.

Man as man's environment: Examples include one man observing another man, human medical observations, human psychological observations, human mythologies, and human genome.

Microscopic environment of man: Examples include biological cells, germs, bacteria, viruses; chemical molecules and clusters; and physical atoms, electrons, photons, neutrinos, quarks, and gluons.

Macroscopic environment of man: Examples include observations of the earth, atmosphere, land, and sea; botanical and zoological observations; observations of the water cycle, the vegetation, the birds, the fish, the minerals, and clouds; and occasional events like lightening, earthquakes, and climate related events such as al Niño. Other examples include galaxies, supernovae, binary stars, the redshift, the universal expansion, and dark matter and energy.

Such observations are analyzed to understand any cause-and-effect relationships between them. This **analysis** helps in understanding relationships between events and entities.

Examples of such analysis include the relation between polar-ice melting and sea-surface rise and its impact on habitability of certain coastal regions; the relationship between carbon-di-oxide emissions and rise in global temperature with an increase in storms and hurricanes; and the relationship between market supply and demand to prices and stock markets and other economic parameters.

Theorization involves a study of correlations, trends, and patterns for the phenomena in the physical world. The study is then generalized using formulated hypotheses. Further experiments, test and analysis are performed that will prove, modify, or disprove a hypothesis. A well-tested hypothesis eventually leads to a theory. The theory is continually subjected to experiments, further observations, and deeper analysis to verify it. As a result, an accepted theory may then be modified, or rejected, and a new theory to replace it may be developed.

Examples of theories include the theory of gravitation; theory of evolution; theory of expanding universe; and theories of global climate change.

Observation, analysis and **theorization** constitute the scientific research process. Henceforth, I will take these as characterizing the scientific research methodology.

Axiom 2

Statement:

Man possesses the innate capability of conscience which connects man with his spiritual world.

Man uses his conscience to know what is his strength and what is his weakness; in other words, conscience knows what is good for man and what is bad for him and informs man what his essence is or what his nature is.

Conscience helps man to live in accord with his essence by heeding his potential and his limitation – to adopt what is good for him and refrain from what is bad for him. Conscience envelops intellect as intellect envelops senses such that conscience helps man to understand the meanings of his intellectual activities just as intellect helps him to understand the meanings of his sensual activities. Therefore, the intellect without the conscience is not very useful just as senses without intellect are not very useful.

Intellect can logically go from one step to another; however, it cannot determine if doing so will be good or bad for man. That is the role of the conscience. Intellect takes man through steps without any inkling which way is bad and which way is good for the life of man. That is where conscience is necessary to guide man towards a happy, fulfilling and peaceful life.

Conscience is a built-in compass for man to steer his life. Innately, conscience remains in touch with the spiritual world. That is how it knows what is good for man and what is bad. When man looks inward that includes, in part, to consult his conscience. Since conscience is always in touch with the spiritual world, consulting the conscience is like opening a window into the spiritual world for making a spiritual observation.

Observations made in the spiritual world are unique in the sense that a person performs these observations on his own self. He is the observer as well as the observed. That is appropriate because the objective is to look inward.

Following are some examples of observations and the entities to be observed while looking inward.

Observations of one's own self: These provide a study of the state of the self at any given stage. Some variables related to the self are happiness, fulfillment and peace. These entities have the same semantics, language wise, but their meanings and realization depend upon the spiritual state of one's own self.

Observations of one's wisdom: These provide the capability to observe, analyze, and theorize in the spiritual world. Just as the senses and the intellect characterize the methodology of the physical sciences, these together with the conscience and the wisdom characterize the methodology of the spiritual sciences. The external manifestation of

33

wisdom is sharing with others. Therefore, a person might observe his sharing practices and correlate them with how they modulate happiness, fulfillment and peace in his life.

Observations of one's altruism: These observations happen after the onset of altruism has been reached. Onset of altruism is indicated when the interpersonal dealings are at least as favorable to others as they are to one's own self. A person can observe the selflessness with which he promotes the happiness and interests of others. He can observe the correlation of his state of altruism with the confidence that other people express in his dealings with them. He can observe how secure people feel their interests are with him and how peaceful they feel in his presence.

Observations of one's oblivion: Oblivion for the moment can be understood as an extreme degree of altruism such that a person is so engrossed in promoting the happiness, fulfillment and peace of others that he becomes oblivious of his own happiness, fulfillment and peace. A person can observe it through an increased depth and breadth of his own altruistic behavior, and an enhanced quality of his own state of happiness, fulfillment and peace.

Such spiritual observations can be analyzed in order to understand the cause-and-effect relationships between spiritual events and entities. Using the equivalence principle, it is possible to use methods of analysis for the spiritual measurements that, to start with, mimic the methods used by the physical sciences.

There is, however, a difficulty in using these scientific tools because of a lack of quantitative estimations for the events and entities in the spiritual world. Generally, there are three main difficulties in analyzing spiritual observations.

First, the spiritual experiences are personal in nature and they are generally not reproducible at will. Such observations can be made only by the person witnessing the spiritual experience, and someone else cannot reproduce those observations because his spiritual state may not be the same.

Second, the nature and reality of the spiritual observations change over the life of the observer, as his spiritual state progresses. Given a different spiritual state, present-day spiritual experience will not resemble the past ones. With changing spiritual state, the observations change, evolve, disappear, or morph into new experiences.

Third, any measurement and analysis techniques used in the spiritual sciences will need to accommodate the above factors.

The spiritual world is governed by overarching principles that are universal. The universality of the experiences in the spiritual world, even though each experience is personal, is similar to the universality of the experiences in the physical world, despite the local nature of all physical experiments. Underlying reason for both universalities is because of the existence of overarching laws in both worlds.

To begin to address measurement and analysis difficulties we can initially take a statistical approach. We consider multiple observations of a similar nature by different people. Similarity of such observations might be indicated, among other things, when different people describe a similar spiritual experience using similar language and imagery. A set of all similar spiritual experiences can be grouped in a similarity class.

A different set of other types of spiritual experiences would lead to another similarity class. All spiritual observations can be analyzed into similarity classes.

The set of all spiritual observations by different people, at different times, and at different levels of spiritual attainment would arrange itself in some taxonomy of similarity classes. By collecting a large enough set of spiritual observations most types of similarity classes can be identified or a critical number of them can be identified. The discovery of the similarity classes and their position in the taxonomy scheme becomes a discipline of research in the spiritual world.

Such a taxonomy scheme for the spiritual observations will span a multidimensional space. Dimensions can include spiritual level of the observer, type of the spiritual experience (similarity class) that the observation describes, type of events that the spiritual experience observes, and type of entities that are involved in the spiritual event, etc. Correlations need to be established between these variables. When such correlations can be established with reasonable assuredness, progress can be made by associating the similarity classes of spiritual observations with certain types of spiritual levels, experiences, entities, and events. Indeed, this process is complex; for example, as the spiritual level advances, the nature of observer's experience with events and entities will change; different observers may use similarly sounding terminology for the experiences, events, and entities, but the level of observer's spiritual attainment will determine the meanings of the terminology. Therefore, the semantics will be complex.

The taxonomy scheme for the spiritual observations is only a tentative determination to be further evaluated and tested. Equivalence principle and the law of shadows (see Axiom 3) are useful for test and validation. A testing program needs to be carried out continuously in order to validate and revalidate the taxonomy scheme and the correlations. The testing regime covers an extended set of observers, an extended duration of the observers' lives, an extended set of observations covering all similarity classes in the taxonomy, and all correlations established within the taxonomy.

The analysis described above would provide a good initial basis for theorization in the spiritual world. When one theorizes the spiritual results, one goes beyond the individual experiences in the spiritual world. One derives conclusions that are held valid across all observers and observations because of the universality of the experiences in the spiritual world.

Application of the toolset of physical sciences to spiritual observations necessarily introduces approximation, symbolism and model forms which are not used by the witnessing process, see Axiom 5 on page 43. Application of this toolset to the observations in spiritual sciences may or may not hold for observations made via witnessing. This is a caution that should be kept in mind, and it emphasizes the need for more research into the nature of witnessing processes.

Axiom 3

Statement:

There exists an equivalence between the physical world and the spiritual world.

The equivalence is such that a set of entities and events in the spiritual world can be translated into an equivalent set of entities and events in the physical world, and vice versa.

Equivalence means that when a set of entities and events are replaced with their equivalent set of entities and events, it will affectively yield equivalent results in the life of a person.

One of the fruits of wisdom is the ability to translate between the physical and the spiritual worlds. This translation establishes an equivalence relation between the contents of the two worlds.

A man is neither just his physical-self nor he is just his spiritual-self. He is a single indivisible reality, not two separate realities. Therefore, man must be able to experience himself as a single indivisible entity, rather than two independent entities, physical and spiritual. The Equivalence Principle is an overarching principle that addresses this aspect regarding unity of man.

The Equivalence Principle allows translating the entities and events in the physical world in terms of the equivalent entities and events in the spiritual world, and vice versa.

This principle becomes evidenced in proportion to the wisdom that one has acquired. The principle remains hidden from the physical sciences because they do not include wisdom within their framework, and therefore miss out on the spiritual world entirely. Wisdom incorporates axioms 4 and 5 which are not included in the toolbox for the physical sciences. Physical science can extend into the spiritual world if its toolbox is enlarged by incorporating axioms 4 and 5. In other words, the toolbox of the physical sciences is incomplete as is. I will have more to say on this subject in the chapter on Spiritual Sciences on page 105.

A man explores the physical and spiritual worlds. He obtains his sensual perceptions and intellectual concepts from two sources: directly from the physical world, and indirectly from the spiritual world via an internal translation performed using wisdom. The results from the two sources need to be equivalent for the sake of internal consistency of man. If the two sets were independent without being equivalent, man would then be confronted with inconsistent sensual feelings and intellectual concepts.

Similarly, results in the spiritual world are obtained in two ways. First, the spiritual entities and events directly observed using wisdom and conscience. Second, via the internal translation of the physical entities and events into the spiritual entities and events. Here the spiritual entities and events derive from two sources: directly from the spiritual world in the first case, and indirectly through an internal translation in the second case. These two sets of the spiritual entities and events must be equivalent, for otherwise man would be confronted with inconsistent spiritual ideas like happiness, fulfillment and peace. Again, equivalence means that the two sets of equivalent entities and events will affectively yield equivalent results in the life of a person.

The Equivalence Principle is necessary for two main reasons. First, the physical and the spiritual worlds in the universe are not independent of each other; rather, the universe is one entity with physical and spiritual attributes. Second, man is an atomic entity not to be viewed as a composition of the physical and the spiritual.

All systems are innately interconnected such that there is only one system in the limit when the wisdom of a person is effectively infinite.

The equivalence between the physical world and the spiritual world means that each effectively contains the other as far as their relationship to man is concerned. The physical world is subtly contained in the spiritual world and the spiritual world is subtly contained in the physical world. The internal communications within man perform a transformation from the spiritual content to the physical content, and vice versa. These transformations are consistent with the nature of life and its internal harmony.

Law of Shadows

The equivalence principle has a corollary, namely the Law of Shadows.

For the Equivalence Principle one works with entities and events that are transformable from the physical world to the spiritual world, and vice versa. The entities and events are objects, or groups of them, that occur in pairs: the inner presence of an object in spiritual world, and the outer presence of the object in physical world. I point out that the transformation mentioned above can be one to many and even many to many.

One of these two objects is regarded as the 'transformed form' of the other. The transformed form is referred to as the shadow. Either of the two objects can be regarded as the shadow of the other. The definition of shadow is symmetric with respect to physical and spiritual worlds. For definiteness, I will regard the entity in the spiritual world as the object, and the corresponding entity in the physical world as its shadow. It can be stated that:

Every object must have a shadow, and every shadow must have an object.

This statement is referred to as the "Law of Shadows".

Again, I remind that depending upon the transformation being one to many or many to many, the object and its shadow can each be represented by a set of things.

Each spiritual accomplishment has an entity in the physical world that it pairs with, so that each spiritual accomplishment has an observable entity as its shadow in the physical world. Similarly, each physical accomplishment has an entity in the spiritual world that it pairs with, so that each physical accomplishment has its counterpart in the spiritual world.

This provides a pathway to experience spirituality as a physically *observable* thing, and hence to *test and validate* it using the techniques of the physical sciences. This is a *breakthrough result* because spirituality otherwise remains unamenable, untestable, and something that cannot be validated or invalidated.

The law of shadows is a great test and validation tool to gauge progress in one's own spiritual journey. It can also be used, with some caveats, to test spiritual claims by others, claims that might be less than bona fide because of the missing observable shadow of a claimed spiritual accomplishment. That is because the Law of Shadows renders it impossible for someone to be spiritual and yet for the spirituality to remain invisible through the excellence of one's external behavior.

Axiom 4

Statement:

Man possesses an innate capability to internalize knowledge.

Man uses internalization to absorb his knowledge to generate wisdom.

Among other things, wisdom enables man to perform internal transformations between physical and spiritual worlds.

Internalization is an innate capability of man that represents the special case where the *observer participates in the observation process*, and owns the outcome of observation within his own self. What he knows from his research and what he does in his life is totally synchronized and compellingly bound together, so that one's own self in consultation with his conscience admits no possibility for its violation.

It is a spiritual happening. It is not a rational conclusion which may be violated when convenient.

Internalization plays a central role in spirituality and spiritual sciences. It is a new element of research methodology to generate wisdom to enrich life. Per the equivalence principle, a counterpart of internalization should also be incorporated in to the physical sciences, and doing so will expand their scope and capabilities. In particular, it will save mankind from its own greed and corruption.

To appreciate the need for expanded physical sciences, let us note that theories and their applications are ends in themselves for the current physical sciences. On the other hand, the spiritual sciences continue those theories and applications into enhancing life itself via the process of internalization. Spirituality continues to hold hand and guide a person to enhance his life, and to enhance the life of his community.

Per the physical sciences man does awful things to himself using scientific theories and their applications. The atomic and nuclear theories told man what destruction a nuclear bomb will cause. Man is not stopped by this knowledge, rather man is incentivized to make this bomb and even use it to bring devastation to the people of Japan. The logic and rationale of the physical science lack wisdom. They allow that man would destroy happiness and peace for mankind.

Industrial processes were made possible by physical sciences and their applications. They caused ruthless pollution due to man's greed supported by science. Mankind, including the scientists, gradually destroyed the planet which was their home, and caused the global climate catastrophe that punishes everybody.

Spiritual sciences do not permit such to happen.

Axiom 5

Statement:

There exists a mode of knowing which is separate and independent from the rational understanding or analytic knowledge. I will call this mode of knowing as 'witnessing'.

Witnessing enables knowing via a vision that demonstrates phenomena in their entirety.

There is another entity that we need to recognize in spiritual sciences. It is called witnessing. It is vital for the spiritual science while it currently plays no role in the physical science. I explain below what witnessing is and why it is necessary.

Consider an indivisible unity that is not composed of any components. Analytical techniques that are in the toolbox of physical science do not apply to it. That is because indivisible unity is an atomic entity, it has no constituent components. One cannot break an indivisible entity into its non-existent components, analyze them using the analytical research methodologies of the physical science, and then put the analysis results together to re-generate an existent indivisible unity. Such a process would re-generate something totally different and inconsistent with the original.

The indivisible unity can therefore not be understood using the methods of analytical research which the physical science employs. A way to 'know' this indivisible unity is to experience it via a vision that demonstrates it at once in its entirety. This is what witnessing is.

Witnessing lets a man seek to know a thing by experiencing it in all its manifestations, without trying to analyze it, or trying to understand it. In this context, understanding and knowing are two different things.

A vision may be ineffable, and not amenable to an explanation. Witnessing is thus different from both the scientific and philosophical approaches which seek to describe it using artifacts such as symbols or modeling forms. It is like experiencing the taste of honey by directly tasting it, verses inferring it from its explanatory description or its chemical or physical characterization.

Witnessing also takes a person farther than the analytical or philosophical research can. Indeed, a time comes when the traveler finds no satisfaction in the scientific or philosophical methods; the traveler yearns for experiential witnessing; yearns for a vision of it. The traveler automatically finds out if he has reached the stage where scientific and philosophical methods no longer satisfy him. At that stage, experiential witnessing is his main source; it is a tool for this stage. It takes a traveler further into spirituality than scientific or philosophical approaches can.

However, this should not underestimate the importance of scientific and philosophical approaches because in the initial stages of a spiritual seeker these are the methods that hold his hand, and they continue to hold his hand until he realizes that they don't and begins to discover new approaches like witnessing.

Comments on Five Axioms

Only the first axiom, regarding the innate senses and the intellect, pertains to the physical world and physical sciences, the remaining four axioms all pertain to the spiritual world and the spiritual sciences.

The first axiom forms the foundation of the physical sciences as well as the spiritual sciences. The toolbox of the physical sciences does not directly include the innate senses and the intellect, but it includes observations, analyses, and theorization which are the results obtained by the application of senses and intellect.

A person's own self has the capability to look inward as well as to look outward. Hence the self knows both the spiritual as well as the physical. A person's own self looks inward and consults the conscience regarding what is good and what is bad for the self. Then the self looks outward and communicates this awareness to the intellect. The second axiom, regarding the innate conscience, allows the intellect to become aware of what is good and what is bad for man.

I note in the parenthesis that Pope Frances describes conscience as 'the most secret core and sanctuary of a person". According to Pope Francis, the voice of the individual's conscience needs to be incorporated into Pastoral discernments, so that apparently similar cases may require different discernments.

In practice, the process is not simple. That is because, as pointed out earlier, the conscience is invariably buried under the influences of the society mores, so that often time when the self tries to consult the conscience, it only manages to consult these society mores, rather than the innate conscience. It takes some spiritual attainment to be able to consult the conscience. In alternative words we say that it needs some wisdom to be able to consult the conscience, and indeed reach the rest of the spiritual world. In the chapter titled Acquiring Wisdom on page 57 I will discuss wisdom in detail; for now, it may be understood as the internalization of knowledge.

Just as the innate senses and intellect are not directly included in the toolbox of the physical sciences, the innate conscience is not directly included in the toolbox of spiritual sciences. Instead, we include wisdom in the spiritual toolbox, which derives from internalization, which in turn incorporates consulting with conscience.

I caution that conscience must not be confused with another concept known as consciousness. While conscience is innate to man, consciousness is not. It represents the awareness that a person acquires using his innate capabilities. A person gets to know entities and events and becomes conscious of them. What populates the awareness domain of a person constitutes his consciousness. Over time, a person forgets that he is aware of certain entities and events though they populate his consciousness. These entities and events are stated to be in a person's subconsciousness. Consciousness is not a spiritual entity though I remark that the content of consciousness includes knowledge that a person has acquired, and this knowledge does play a crucial role in my theory of spirituality; knowledge is a component of wisdom, which is a spiritual entity.

The third axiom, namely the equivalence principle, establishes an equivalence or even some kind of unification between the spiritual and physical worlds. One could visualize it as a single unified universe that has a physical projection and a spiritual projection. These projections are created by the interface of man with the universe respectively as man looks outward or inward.

Universe with two equivalent projections is a unique feature of this theory. None of the other discourses assert an explicit relationship between the physical and the spiritual worlds; in particular, none asserts such an equivalence between them.

Existence of an equivalence affords a way to visualize spiritual experiences in terms of equivalent physical aspects such that the spiritual experiences can be estimated and evaluated in terms of physical aspects using physical science. This is another unique feature of my theory because no other discourse provides a method of evaluation and validation for spiritual experiences.

It is significant for the life of a person that he sees his life as a unified whole; that he does not partitions his life into spiritual and physical compartments. Partitioning will polarize his life and reduce his happiness, fulfillment and peace.

The fourth axiom, regarding the internalization of knowledge, is an essential ingredient of spirituality. Without it no spiritual experience is possible. That is because internalization generates wisdom which is another name for spirituality. It stands for a human character that is identical inside and out; a character that appears outwardly the same way as it really is. It has no pretention, no greed and no deceit; it represents a purity that desires benefits and safeguards for everyone at least as passionately as it does for one's own self. This could be regarded as the proverb to do unto others as you would want them to do unto you, though the proverb is a very weak expression of what internalization actually produces.

To illustrate the point, consider the fact that good food is nourishing and the quality of fire is that it burns. Having known this knowledge, its internalization implies that you will act according to what you know; for your own self you will desire good food and safeguard from fire.

This much is easy to understand. You will expect that even without internalization. What about the next step, namely that you will desire good food and safeguard from fire for all other people? Logically speaking or on rational grounds this result does not follow from the physical sciences: humans do not commit to providing everyone good food and safeguarding everyone from fire. Though there exist hand waving suggestions that doing so is good for everyone, humans do not really act as if they mean to commit to it; that is why we make atom bomb and drop it on others; and that is why we pollute the earth and cause global climate catastrophe for mankind.

Because of internalization, a person gains a threshold of wisdom to be able to consult his conscience. You can ask if it is good for my own self to desire good food for others, or should it be the other way around. The conscience knows what is good for you, and answers that you should desire good food for others. Similarly, the conscience answers that you should safeguard others from fire. This is how one reaches the threshold result that you desire good food and safeguard from fire for your own self as well as for others. Lest someone be mistaken that this is a volunteer attitude; internalization affectively generates honesty, commitment and compassion which leave no choice for a person to violate what he knows within yourself. The conscience of a person plays a critical role in the process.

As asserted earlier, the conscience is usually buried under the influences of the society, and you may not get the affirmation that you should desire good food for others, and protect them from fire. In that case, you may need to travel some more along the spiritual journey before you can access your conscience.

With increasing wisdom, internalization can and will demand more. It will demand altruism from the one who practices internalization; moreover, it will go further and demand oblivion in the sense that one forgets about benefits and safeguards for one's own self, and devotes to the benefits and safeguards for others. These are different levels of practicing internalization with varying levels of wisdom; I will discuss oblivion further in the chapter titled Stages of Oblivion on page 65.

The fifth axiom, regarding the witnessing, is about a mode of knowing things without analysis and rational approaches. Witnessing is like an explicit demonstration or vision of a phenomenon. In this mode one knows the phenomenon without understanding it in terms of analysis or rationality. This way of knowing does not currently exist in the toolbox of physical sciences.

Witnessing is essential in knowing things that are an indivisible unity, because in such cases analytic techniques do not apply. However, the application of witnessing is not limited to such cases; things that can be analyzed can also become known through witnessing. In such cases, witnessing is an alternative, though not necessarily equivalent, way to know phenomena.

The toolbox for spiritual sciences includes observations analyses and theorization in common with the toolbox for physical sciences and it also includes internalization and witnessing which are two additions to the toolbox. Thus, the spiritual sciences embed the physical sciences within them, so that all spiritual phenomena will be in compliance with the requirements of the physical sciences. This does not refute the possibility of unusual things, though it does assert that such things will occur within the laws.

Meaning of Self

Currently the concept of "self" is unclear. It is often thought of as ego, which in turn is perceived as self-perception, which does not represent the true "self" as a noumenon. Hindus approach the self as an illusion because they are driven by their religious notions. This situation adds further non clarity to the understanding of "self". Because of this lack of clarity, people add adjectives such as inner self or true self, which does nothing to clarify and merely adds to the list of terms needing clarification.

Related religion inspired concepts also exist, like soul in the West and Atman in the East. Roughly speaking they mean the inner most essence of a person, which sounds good but fails to communicate anything specific. Such concepts are vague; therefore, I do not use them in this treatise.

In this treatise, I will use the concept of "self" as a representation of an individual person to signify the innateness in him and its derivatives. I will often use words like 'a person's own self' to emphasize its role as representing an individual. Self consists of all things innate in a person, as well as things acquired through their use. As such it includes physical as well as spiritual aspects of a person and possesses a spiritual nature and a physical nature. The two together constitute man's nature. Man prospers and his life is happy, fulfilled and peaceful when he lives true to his nature.

Among other things, such a self may communicate with spiritual things like a person's innate conscience. This happens provided that the self is sufficiently pure from greed, corruption and society mores which intervene between the self of a person and his conscience. Alternatively, we will describe this situation by saying that a person's conscience is buried under greed, corruption and society mores. In such situations, a person's conscience can become inaccessible to his own self.

This concept of the self that I have introduced is well defined and ready for further exposition and research. It is not vague like Soul or Atman. Self is immanent and fully represents the essence of a person, both spiritual and physical. It provides means to consult with the conscience, which guides man regarding his strengths and weaknesses, and opens a window into the spiritual world so that a person's self can make spiritual observations that make him wise and ready for proportionately deeper spiritual experiences.

When the self is pure enough to be able to consult the conscience, such a consultation is a dialogue within a person, it is a mode of looking inward and to discover spirituality. This is how the spiritual journey of a person begins, namely with a consultation with his conscience. However, this is only a beginning, there would be many such consultations. Such consultations with conscience let the self know what its nature is, what its true potential is, where the pitfalls lie, what things are beneficial for it and what are harmful. This happens gradually over time, event by event. It is not like preaching by the conscience, nor it is an information store. Each time there is an incidence that the self decides it needs consultation with the conscience, a consultation takes place. It happens by a process that is spiritual, and has its prerequisites and scope.

Having known its nature and its potential, the self is able to move along its spiritual journey. It can experiment with its own nature to discover it more deeply and widely. This initiative is by the self after having the benefit of consultation with the conscience. Another way of expressing this experiment is to say that the conscience has opened a window for the self. The self reaches out farther into the spiritual world through this window. This is how the self travels on the spiritual spiral, moving higher in its stages and reaching out wider in scope.

Thus, all activity, spiritual and physical, starts with the self of a person, and grows with the self. The self includes the use of senses, intellect, conscience, internalization, and witnessing, as well as the state of wisdom, sharing, altruism and oblivion. As these things progress the self grows and with it grow the nature and impact of all activity. Therefore, the self is dynamic and reflective.

Acquiring Wisdom

Knowledge and internalization are vital ingredients of spiritual life. Knowledge is initially acquired in response to the physical needs of existence. It is useful in meeting those needs when it is adequately applied. Among other things, the adequacy of application depends on the level of internalization of the knowledge. Internalization minimally requires and enforces that man desires the wellbeing and happiness of others as compassionately as he does his own wellbeing and happiness.

I have defined the concept of wisdom to help understand the combined dynamics of knowledge and internalization. I define wisdom as a multiplicative product of knowledge and internalization.

$$(Wisdom) = (Knowledge) \times (Internalization)$$

This definition is notional because exact measurements for knowledge and internalization are not yet defined at this stage of research. Many scientists possess a lot of knowledge but often they lack in its internalization, so their wisdom is low. As a result, they help produce wonderful inventions and yet also help their uses that are not humane or are corrupt. That is how we have weapons of mass destruction and practices that lead to things like global climate catastrophe.

On the other hand, some dogmatic persons often have little knowledge but they internalize it to a high degree. Their wisdom is also low.

Wisdom is required for a happy, fulfilled and peaceful life. It represents how well man knows his own essence in relationship to the universe around him. A man's wisdom represents the extent to which he has discovered and known his own self. It also represents his spiritual state because self-discovery is equivalent to spiritual maturity.

Among other things, wisdom is used to transform spiritual aspects into physical aspects, and vice versa, per the equivalence principle.

Wisdom is an acquired capability. It is not innate to man. However, it derives from the exercise of senses, intellect, conscience, and internalization which are innate to man. In order to become wise a man must exercise and enhance all of his innate faculties.

As the wisdom of a person increases things start to happen in his spiritual world, and they evolve and grow as his wisdom grows. That is why the state of wisdom of a person is synonymous with his spiritual state.

When the wisdom reaches a threshold, a person can reach out and consult with his innate conscience. This allows a person to know his own essence, his strengths and his weaknesses. When wisdom reaches the next threshold, a person starts to understand altruism and act accordingly. Such a person realizes the value of service to others and devotes to it. He also shares his possessions with others without expecting anything in return, achieving advanced inter personal conduct.

The person keeps growing in the depth and breadth of his wisdom. The next threshold is for oblivion, when the person is so engrossed in service to others that he forgets about serving his own wellbeing. He realizes that others are part of himself and when he serves them, he also serves himself. Oblivion sets in with this realization and acting accordingly. At this stage, the person begins to feel the unity of existence. This state continues to intensify, and he actually forgets about his own separate existence. The sequence continues and he keeps experiencing the wonders and ecstasy of a progressively unified existence.

A sequence also exists in the negative direction. When a person falls below the first threshold, he loses the ability to consult his own conscience. He loses the realization of his own essence, does not know his own worth or his limitations, and confuses between good and bad. When he falls below the next lower threshold, he becomes greedy and seeks after possessions at the expense of others. He loses human compassion and becomes cruel. The sequence continues lower and lower, and he demonizes himself more and more.

Sharing Practices

Sharing with other people is a spiritual activity required by wisdom. It is a specialized concept in my theory. It is totally different from the common practice of charity.

A person shares his happiness and peace with others because he regards the happiness and peace of others as his own. Sharing enhances everyone's happiness and peace.

A person can share with others his possessions, his time and his wisdom. For that purpose, people do various activities like volunteering at schools and hospitals, providing food and shelter to the homeless, and joining activism to improve the society in some aspect of living. All types of sharing activities are fine so long as they make the lives of others happy, and the person finds that the activities continually enhance his own wisdom and his own spiritual experience. Both these aspects are necessary for an authentic sharing activity.

A person must be genuine in his thinking and actions towards enhancing the happiness and interests of other people. Other people include everyone, from his beloved ones to the man in the street, to those who have animosity towards him. Gradually he achieves sharing with a larger and larger number of people, and to a greater extent and meaningfulness. Sharing must be achieved in a way that is in line with the wisdom of a person and his spiritual experience. That is because his own self validates his behavior, in consultation with his innate conscience.

He shares what he has with others, and lets others share with him what they have. Objective of sharing is not to be charitable or to be a wise man to others. Rather, the parties in the sharing activity are at parity. The objective is to share spiritual experiences and enhance spiritual experiences for all by sharing in a manner that is respectful and altruistic and even oblivious. Sharing possessions is not charity, and sharing wisdom is not preaching. There is no upper and lower hand because when you share with others, others share with you, there is parity between all. If you give you must also accept, and if you accept you must also give, that is what parity implies. You must give genuinely even if giving is only sharing a smile or a kind compassionate word.

Sharing at large scale, with all people, whether friends or foes, and with depths of feelings and commitment, produces compassion. It generates tendency to internalize knowledge such that you act compassionately and in accord with what you know within yourself. In other words, you act wisely.

Excellence in sharing practices is a reflection of your wisdom, and your wisdom requires proportionate excellence in your sharing practices. Wisdom and sharing practices are reflections of each other. By improving the excellence in your sharing practices, you improve in wisdom. Conversely, when you improve in wisdom you also improve in the excellence of your sharing practices. Sharing practices and wisdom are shadows of each other: one enhances the other. More obliviously one shares, more he advances in wisdom. Conversely, wiser a person is, more obliviously he shares. More wisdom a person acquires more he dives into oblivion, and closer he gets to everybody, until he can literally share and feel their pain and pleasure. As he advances in his wisdom he shares in more advanced ways, he gets ever closer to others, and he can feel their pain and pleasure ever more intensely and more personally. When he gets very close, proportionately his pain and pleasure and their pain and pleasure become one and the same thing.

The difference gradually disappears and the identity dissolves into a unified existence. He becomes them and they become him. This result obtains when the wisdom increases greatly, and also when the excellence in sharing practices increases greatly. The two aspects are two equivalent attributes of a person.

Stages of Oblivion

Oblivion in the present context means forgetting the benefits and safety of one's own self in an enhanced desire to serve the benefits and safety for others, knowing that doing so will also safeguard the benefits and safety of one's own self. It is a state of being beyond altruism.

Oblivion is not binary which either exists or does not exist. Rather, like wisdom it is progressive, always reaching higher experientially, and wider scope wise, like a rising and expanding spiral. As a person dwells deeper and deeper into oblivion, he gradually changes from a primarily physical person to a primarily spiritual person. The laws of nature that operate in the spiritual world take him there, even if he is unaware of those laws. It is rather like the force of gravity in the physical world that guides a river from its source to its delta, even though the river is unaware of gravity.

Lest someone thinks that it is a matter of volunteer attitude whose calculations change when expedient to do so. It is not the result of a rational deliberation. It is a necessary and spiritual consequence of the stage of wisdom of a person and the corresponding stage of his oblivion. Just as two plus two equal four, that stage of wisdom equals oblivion where one necessarily knows that one's own happiness and peace is embedded in the happiness and peace of everyone else. It is an outcome of the spiritual laws that make the result so compelling that no choice is left to be otherwise.

More spiritual a person is, deeper he dwells into oblivion. As markers, the progression goes through sharing, altruism, oblivion, and to a unified existence. Altruism begins when a person cares for the wellbeing of others at least as much, as he cares for his own wellbeing. A person is driven to this result by his state of wisdom, which compellingly leads to this conclusion without an option to be otherwise. It is not because the options are eliminated, it is just that the one course that the oblivion shows is compelling.

As he travels this journey, the meaning of a spiritual living keeps changing. Initially, a small level of spirituality in a person would suffice his life. However, as he excels in wisdom, he is no longer satisfied with a small level of spirituality, because the nature of happiness, fulfillment and peace that he seeks has changed. Going deeper into oblivion will even change the meaning of wisdom itself. It will change the nature of his yearning for wisdom. New doors are opened for him that he could not even imagine existed. He sees new ways to know and experience wisdom, and consequently new experiences of oblivion. Thus, it goes spiraling upwards, and expanding sideways, into higher and higher spiritual experiences growing in their intensity, their nature and their scope. The spiritual spiral never stops; it stretches infinitely into higher and higher levels of spirituality, and it spreads in its scope to encompass everything. Such is realized by knowing that one exists only as part of everything, part of a unified existence.

As the wisdom increases, a person realizes, more and more, that he does not exist by himself; rather he exists as part of everyone else's existence. He realizes that everybody else too does not exist by themselves; rather each person exists as part of a bigger existence produced collectively by all, a collective existence, a unified existence.

As this state of oblivion progresses, his actions morph into a refined and advanced form where he acts out of an inner compulsion induced by his spiritual state. Again, I remind that this happens as a dynamical result under the spiritual laws, such that the person does not reach there as a matter of choice but as a matter of necessarily derived spiritual consequence.

In the limit when the wisdom of a person approaches infinity, he emerges as a purely spiritual entity. He encompasses the nature of everything. The opposites join, differences annihilate, time and space cease to exist, and the scientific understanding that is based on time and space dissolves into a vision of spiritual witnessing. Many such things do not lend themselves well to verbalization, they can be witnessed but are ineffable.

I would like to note that oblivion is a spiritual progression with increasing wisdom, that takes place according to the well-established laws in the spiritual world. It is not like annihilation of the 'self', it is more like the 'self' meeting its bigger vision, ultimately an infinitely bigger vision.

Prerequisites for Spiritual Beginners

How can a man begin the spiritual journey and achieve happiness, fulfillment and peace? Man needs the following three prerequisites.

Trust your own self

Generally, a person trusts the people that he knows. He trusts them more than those that he does not know. The person he knows the most is his own self. Hence, it makes sense that he should trust his own self the most.

What does it mean for a person to trust his own self? It means that, as an adult, he makes all his life decisions himself. In the process, he can consult other people, he can read books, and he can watch what happens within himself and within his environment. Find out facts that he can trust, and make the decisions of his life using those facts in the light of what his own self tells him. It is necessary that a person learn to listen to his own self, and trust his own self.

In the process, he can sometimes make wrong decisions but the process will lead him to realize his mistake and will offer him an opportunity to correct it. Making mistakes is part of living, without exceptions. If mistakes must be made in life, it is better that a person makes them himself, so that he can own up to them and be able to recognize them in due course to correct them.

If a person lets other people make the mistakes for his life, then he might not know how to detect them and subsequently correct them.

Moreover, other people may be more willing and casual about making mistakes for his life. After all, it is his life and they do not have to live the consequences of those mistakes. More importantly, other people do not know his own self; they do not know his passions and his dreams, they do not know what brings him peace and fulfillment. If he trusts his life's decisions to others, would he feel that he is fully living his life? How likely is he, in such a scenario, to feel happy, fulfilled and at peace?

I have so far assumed that the authority that a person trusted with his life's decisions was well meaning and reasonably competent. What will happen if the authority gained his trust with a malicious intent, or under some ulterior motive? Even if the authority was not malicious, what will happen if the authority was incompetent? Therefore, a person must trust his own self with the decisions of his life.

There is a need for a person to make optimal decisions in his life and to minimize the mistakes that he might make. For that purpose, it is necessary that he discover his own self as well as his environment. These two discoveries are a lifelong process. They make a person wise. More discoveries he makes the wiser he becomes. With wisdom, he can make near optimum decisions for his life, and eliminate many mistakes. Wisdom is the key to a life that is happy, fulfilled and peaceful. Wisdom gives him a global picture of his life. It shows him what resonates with his own self and what does not. It gives him a purpose according to which he decides if a particular decision is collinear with the purpose or not. Wisdom is a compass to know in which direction he wants to head and where his next stop should be on the way to greater happiness, fulfilment, peace, and enlightenment.

Keep an Open Mind

An open mind requires the following three things: a person takes a position conscientiously in resonance with his own self, versus a position based on external considerations and ideas; a person should be prepared to revise his position when new evidence presents itself, versus force fitting the new evidence into rigid or dogmatic positions; and a person should actively search and re-search for data and evidence that might necessitate a revision of his position.

An open mind sets a person free because he is free to think according to his own self and to take actions that resonate with his own self. In particular, he has no obligation to pledge allegiance to external consideration and ideas with which his own self does not resonate.

A closed mind will enslave a person to rigid or dogmatic positions, and make it difficult if not impossible to revise or change this position. When new facts come to evidence, a closed mind will deform the evidence to make it fit a dogmatic position. A closed mind is constrained within a bubble, and its dogmatic actions constrain his own self. This inner conflict generally produces unhappiness, distress, and disappointment. On the other hand, a closed mind can be so closed that it will dismiss any evidence against his closely held dogmas without even considering it; then it might remain in its bubble without feeling inner conflicts, and thus avoid unhappiness, though without tasting genuine happiness. Genuine happiness is dynamic; it evolves and grows. It is possible to experience this growth through feelings of resonance with his own self. For a dogmatic life too, there can be a feeling of resonance, not with his own self, but with an externally

71

induced ideal which can provide a contrived happiness that feel static, without evolution and growth.

How does a person know if he is keeping an open mind? The fruits of his thoughts and actions will reveal that. The fruits will produce happiness, fulfillment and peace if his mind is open, He has an open mind if he can think and act without ad hoc limitations constraining him within a box, if he can act within the full scope that the reasoning permits, and if his thoughts and actions are in harmony with his own self.

An open mind can still err in its thoughts and actions because of the limited wisdom of a person, but an open mind will inevitably detect and correct such errors in due course.

Search and re-search for a clean worldview

Only an open mind will let a person honestly search for the facts that might challenge his own positions and values that he holds in his worldview.

Search for facts that might challenge one's values helps to cleanse one's worldview by recognizing the false values that a person has acquired and then to purge them. Presence of false values in the thoughts and actions will suppress the voices of one's own self and consequently the ability to consult one's innate conscience. To progress spiritually a prerequisite is to cleanse oneself of the false values. It is for this reason that a person must search and re-search.

A person might wonder what the false values are and how to recognize them, and how to cleanse himself of these values. Recall the litmus test: the false values will reduce the inner happiness, they will gradually destroy the inner peace, and they will erode inner fulfillment. If a person is repeatedly disappointed while he tries his best to achieve happiness, fulfillment and peace, then his roadmap is based on false values. He should take a pause, reexamine what he is thinking and doing, search for the false values, identify them, and drop them from his worldview. It is a difficult task that takes long experiments but it needs to get done for happiness, fulfillment and peace.

Values determine the validity of actions. Actions motivated by false values are invalid; they will only reduce one's happiness, fulfillment and peace. Usual criteria people use include such goals as success. In many societies, the word success equates with being rich in the sense that a person has a lot of possessions that money can buy in his society. On his way to success, he finds a job and climbs the corporate ladder, or he does business and climbs the ladder of fortune, or he becomes a politician and climbs the ladder of electoral competitiveness, or he becomes a church minister and climbs the ladder to becoming the chief priest. When he climbs such ladders, his society admires him and regards him as successful. All these things are external to him. They do not make a call on his own self; rather, they deform and mold his own self to conform to externally formulated success criteria. Is there then a surprise that these things might not lead him to feel internally happy, fulfilled, and peaceful? That is because his own self does not resonate with these externally imposed values; and his innate conscience does not resonate with them.

A value is purged as false if and only if one's own self rejects it. Therefore, a person searches for values that resonate with his own self, and adopts them. One uses these values to redraw the map for travelling the journey of his life. The effort to weed out the false values and test out the valid values is the embryo of spiritual progress towards progressively greater happiness, fulfillment and peace.

It is not enough to go through this search process once. A person needs to come back to it repeatedly. He searches and re-searches. That is because he is liable to make errors about his value system so long as some false values still occupy his thoughts and actions. He comes back and reexamines and revalidates his values under many circumstances including the following two: when new evidence comes to his knowledge that he feels might lead him to a different decision with respect to the reasoning to reject or keep a particular value; and when his own self tells him that he is unhappy, distressed, or unfulfilled in his life, which is the time to take a pause and reexamine the values in his worldview. An open mind empowers a person to do genuine search and re-search, which is a bases for spiritual progress.

How to Look Inward

The prerequisites purify the worldview of a person for his spiritual seeking. Purification of the worldview means that he purges the false values in his worldview and replaces them with valid values that his own self resonates with, his conscience approves of, his search for the facts validates, and his re-search continually reexamines for validation or invalidation. During this process he trusts his own self as his guide.

With such purification a person's own self is wise enough to consult with his innate conscience. It is in this sense that I sometimes regard the self and the conscience as synonym. A person's own self is purified enough that the conscience will inform him regarding what is good for him and what is bad for him. In particular, he will know which direction will take him to peacefulness and which direction will take him where distress and depression awaits.

Once a person is in consultation with his innate conscience, he is also able to exercise his other innate capabilities in a way that can lead him to experience additional spiritual entities and events. In other words, a window opens for him to make spiritual observations. I will call this way of making spiritual observations as recalling the innateness. It can superficially resemble the mindfulness yoga, but is actually very different.

Therefore, a person's own self does more than consult with the conscience; it can go beyond into the spiritual world in proportion to the wisdom. This happens generally when the self consults with the conscience which opens a window for the self into the spiritual world so that it can make more extensive observations of spiritual entities and events, in accord with the state of wisdom; in this sense, the state of wisdom is synonym with the spiritual state.

I refer to the above process as 'recalling the innateness'. It resembles mindfulness in that it requires non-distracting and soothing environment. A person lets his body and his breathing relax in a posture that naturally supports them. The idea is for him to be completely with his own self, avoiding environment that can distract him outward. During this, it is very helpful if the person is in Nature instead of being in manmade artificial environment. Nature by itself is an inducement to spirituality so that being in Nature facilitates the process of recall to innateness which is generally of two types.

In one type a person sits silently and tries not to think at all. It empties his mind of the external for that duration of time. When he empties himself, he becomes proportionately free of the external influences. Freeing from external influences is conducive to looking inward and thus to learn to perceive his innate spirituality directly. He can become aware of his innateness and recall it to experience oneness with it. In this scenario mindfulness prepares for the recall to happen. However, recalling innateness is a separate and independent spiritual event which does not happen as a result of mindfulness exercise. Separate spiritual seeking is required for recalling innateness.

The second type of recall to innateness is a form of heightened concentration. Again, a person sits in the same kind of Natural environment, and focuses his attention on seeking answer to a specific question, physical or spiritual. Note that understanding the question is part of the answer. He is more likely to arrive at an answer during the recall to innateness because he is undistracted and wholly focused on that one question. For example, he may meditate on his life, whether he is adequately happy, fulfilled and peaceful.

He can concentrate on listening to his own self. It may tell him about his wisdom, his internalization, and his conscience. It may remind him of his sharing practices; those he kept and those he failed to keep. It may ask him to acknowledge his errors, for example, when he missed an opportunity for sharing with others. It may provide him with new insights that enhance his sharing practices.

One's own innate conscience is always there speaking to him, even during the mundane actions and events of life. It is up to his self to listen to it. The ability to listen is according to the state of purification of the worldview, or equivalently the state of wisdom.

A person can listen to how his own conscience appraises his spiritual state, versus what he thinks his spiritual state is. Is he as spiritual as he thinks? Is he as spiritual as he wants to be? What is the gap between his actual behavior and the required behavior for his aspired spiritual state? How does he fill the gaps? How does he get ready for the next spiritual stage, even though he may not be aware of what is coming? What are the potential hindrances to the next stage? How will he navigate these hurdles?

He examines his own emotions and feelings, if they are positively or negatively oriented, and if they support or hinder his quest for happiness and peace. How naturally his decisions and actions flow, do they bring him happiness and peace? How does he feel about others around him, and the world that surrounds him; is it a feeling of peace and tranquility? Where is he along the spiral of his spiritual journey, for example is he wise, how are his sharing practices, is he altruistic, does he feel oblivion? These things are part of his appraisal of his spiritual state during his recall to innateness.

A person does this appraisal on his own because it is primarily an exercise at looking inward. However, I demonstrated earlier that this is a spiritual science with laws that apply to the spiritual world to which our individual spirituality belongs. There is advantage to also do it together with others and share the experiences in the spirit of a sharing practice. For a beginner, it is non distracting to do so with people of like inclinations.

Spiritual Tools

In order to share, a person needs to interact with other people. Social occasions can provide a mean to interact with other people in a sharing spirit. Some rituals can also be affective for a sharing interaction among people. Social customs and rituals can therefore be important in the context of sharing. That is because sharing requires physical actions between people for its manifestations.

Many rituals and customs have existed in widely different cultures in history. Some of these may be rooted in myths and superstition; others may be rooted in the wisdom, experience and cumulative knowledge that have been validated. Therefore, rituals and customs can have widely different meanings and widely different connotations. If a person does use rituals as tools, he should remember to do it according to his own wisdom and his own experience of spirituality. The objective in using rituals is to enhance his experience of wisdom and spirituality. If they help him towards that end, he can use them; if they do not help him, he can pass them. A person should test them out for validity using his current state of wisdom and spirituality.

Some of the well-established rituals include prayer and pilgrimage.

Ritualistic prayers can often be useful as a tool to experience spirituality. During such prayers, a person can concentrate on his own self, to reflect on the current state of his wisdom, sharing practices, and how he can improve in wisdom and sharing. During ritualistic congregational prayers he can concentrate on his dealings with others, on the affairs of the world, and conditions of man.

However, it is important for a person to evaluate the usefulness of the particular ritualistic prayer in which he intends to partake. This evaluation needs to happen before, during, and after he partakes in a ritualistic prayer. This evaluation may include specific questions. Does the ritualistic prayer help him in his spirituality? Does it enhance his thoughts and concentration? Does it make his sharing practices more meaningful? Such evaluation depends upon his current spiritual state, and therefore needs to be redone over time, as the spiritual state evolves.

Besides the ritualistic prayer I propose what I call prayer of the heart. It means that a person has a deep conversation with his own self. Included in this conversation are his desires and efforts. He scrutinizes his desires and passions to assess how well they represent his own self. He scrutinizes his efforts to assess how well they reflect his wisdom.

To have desires is natural to man, and to crush desires is not natural or helpful. Desires provide motivation for a man's actions and help life flow forward. How to have desires but no sadness or pain? On the surface the idea seems to be inconsistent with Buddhist thought which seems to imply that desires will lead to pain. If the desires do not come true, does that necessarily lead to pain or sadness? It depends upon the worldview of a person. It is important to realize that there are external factors that we do not determine or control, and yet these factors are crucial for our desires to come true. While it is fine to desire, it is not right to insist that the desires be fulfilled, because the external forces might not align, and we must not make our happiness a hostage to the external elements. We should detach our desires from our happiness. We can desire with the realization that our happiness does not critically depend upon their fulfillment. Our happiness is in our own self, and the achievement of our happiness must remain in our own hands.

Pilgrimage can be useful to acquire knowledge, enhance wisdom, and practice sharing. In some cases, the pilgrims come from all over the globe and they dress alike while carrying out the rituals. Such situations bring forth the intrinsic equality of all men, and remove the artifacts of the society that distinguish between men for their social status. The sharing experience in such situations can be inspirational; for example, a poor man may be charitable to a rich man because in such settings of equality neither knows the socioeconomic status of the other.

Most pilgrimage involves travel. Travel can be revealing and enlightening. It can help dispel preconceived notions. The indoctrinations of a society can become more apparent when a person travels and exposes himself to other societies, traditions, and cultures. Thus, he can cleanse himself of the indoctrinations after they become apparent to him. He can also visit historical places and learn how the people have experienced spirituality over historical periods.

A person needs to assess a particular pilgrimage carefully before he goes on one. In order to be acceptable, as a minimum it has to fit into his current spiritual state. Pilgrimage offers the advantage that he travels in the company of other people who are also seeking spirituality. This brings to him a concrete situation of how other people perceive of spirituality and how they experience it. If he performs multiple pilgrimages under multiple perspectives, he can contextualize his own experience of spirituality versus that of other people who may view it differently. This can broaden his worldview and his appreciation of what spirituality can be; perhaps make him see that his own self is, in essence, the self in all.

Spiritual Journey: An Illustration

The spiritual journey of a person is not so much in envisioning it as it is in travelling it. I will present some reflections that may serve as signposts along this journey. They are merely pointers to possible occurrences, although each journey is unique and its experiences are specialized for the traveler.

Nature of the universe

Things exist in the physical world and in the spiritual world, as revealed respectively by the outward-looking and inward-looking views of man. These are interconnected by the equivalence principle, which roughly means that the spiritual world can be transformed into the physical world, and vice versa. Further, these things are paired such that they act as if one of them is the shadow of the other; though this pairing is likely one to many or even many to many. A thing cannot exist without its shadow just as a shadow cannot exist without the thing.

The physical and spiritual worlds are projections of the universe. If a thing exists in one projection, a corresponding thing, or a set of those, must also exist in the other projection. Further, since inward looking and outward looking are exhaustive there is nothing in the universe other than the physical world and the spiritual world.

Man and Universe

Man relates to the universe through his innate capabilities: the senses, the intellect, the conscience, the internalization and the witnessing.

These are man's interfaces with the physical and spiritual worlds through which man explores these two worlds.

Knowing one's own self

Man uses his innate senses and intellect to discover the physical world, his innate conscience to discover the spiritual world, and his acquired wisdom to translate between them. Man acquires wisdom via the exercise of his innate capabilities. It represents how well man knows his own self. Knowing one's own self is the only way for man to become spiritual and to lead a happy, peaceful, and fulfilled life.

Wisdom shows man the spirituality of the physical and the physicality of the spiritual. More wisdom a person has acquired, more clearly he can see the unity among the physical and the spiritual. The two are intertwined and entangled.

The full wisdom of the universe resides only in the entire universe, it cannot reside in a part of the universe. Therefore, man cannot reach it as an individual entity, though he can reach it as part of the entire existence, namely when man as an entity is dissolved into the unified existence.

Wisdom is the criterion for the spiritual journey, which is also the journey of knowing one's own self. As his wisdom increases, man goes deeper into the spiritual world and he reaches higher echelons of happiness, fulfilment and peace. As his wisdom increases, a person experiences changes within himself and undergoes a transformation. These changes and transformation are not hidden as the content of his spirituality, on the other hand they are amply reflected as they manifest in his outer behavior. For example, they are visible in the excellence of his behavior with others, in his style and mannerism, in his extent of sharing with others, in the altruism in his behavior, and in his extent of oblivion where he serves others and forgets about himself.

Journey and some milestones

In order to progress through a spiritual journey, the core requirement is the wisdom, which begins when knowledge is internalized. Following practical steps can prove helpful in the spiritual journey.

Healthy body: A person needs to live in healthy environment. He needs to breathe clean air, drink unpolluted water, eat natural foods, and live in bright airy quarters.

Healthy mind: A person must possess the following three prerequisites for a healthy mind: trusting oneself, maintaining an open mind, and searching and re-searching for the facts.

Healthy worldview: For a healthy worldview man must enhance his prerequisites, exercise his innate faculties to recognize dogmas in his thinking and actions, and disown any dogmas. Some practical steps can help purify his worldview, for example recall to innateness, ritual prayer and pilgrimage can be helpful as well as the prayer of the heart.

Healthy sharing: Enhanced wisdom implies enhanced sharing practices. Sharing has two main aspects, sharing of wisdom and sharing of possessions. Sharing of possessions brings a person closer to those with whom he shares. It produces a bond between them that unifies them proportionately.

Sharing of wisdom happens genuinely after a man encounters the equivalence principle. Among other things, it produces wisdom about wisdom. Many aspects of sharing wisdom look like teaching, often not traditionally but in ineffable ways.

Altruism: A person enters altruism when he cares about the happiness and interests of others at least as much as he cares for his own. This starts as a person begins to successfully communicate with his conscience. As the conscience opens for him more and more windows into the spiritual world, the experience of altruism deepens and broadens. Altruism gradually begins to look as though the person cares less and less about his own happiness and interests, and he begins to care more and more about the happiness and interests of others. This is when altruism begins to merge into oblivion.

Oblivion: At the stage when a person enters altruism, his wisdom has demonstrated to him that serving others is the same thing as serving his own self. As a person enters further into oblivion, he realizes that there is no real difference between himself and others. He realizes that a separate and independent existence is a mirage. This is the beginning of the individual identity dissolving away. Oblivion sets in when a person is unaware of any differences between his existence and the existence of others, because he tends to experience all existence as one. This lack of awareness is not due to some fuzziness of understanding or fogginess of the worldview. It arises from an illumination of his own self such that a person becomes acutely aware that his own essence can be realized only as a part of unified existence. Oblivion is an infinite region of the spiritual world that a person explores on his own. When someone reaches there, he refrains from an attempt to verbalize it. He can share it somewhat with those who are in that vicinity.

However, there are others who have tried to express it, either as their own experience or what they heard or imagined about it.

Validation: Throughout the journey of getting to know one's own self, the traveler must validate his state. According to the law of shadows, a person can validate his state by scrutinizing the excellence in his inter personal behavior and his sharing practices.

Everyone goes through this journey of getting to know one's own self in his own way. There is no universal prescription for it, nor are there universal milestones. Each journey is unique. There are as many paths as the travelers. The only metric along this path consists of meeting the three prerequisites, and how the innate faculties are used. Otherwise, there are no universal rights and wrongs along this journey, nor are there universal do and don't instructions. Everyone feels his way and progresses along the spiritual spiral at his own pace and along his own path.

The final milestone in this list is discussed below.

World Peace and Prosperity: Final Milestone

This is a new concept in spirituality that my theory introduces and facilitates. Traditional spirituality is personal and does not regard community happiness or world peace as being within its scope. My theory thus fills a void that traditional spirituality leaves behind.

How does spirituality apply to a community? How does a happy, peaceful, and prosperous community arise?

The discourse on spirituality so far has limited itself to man as an individual. For an entire community, I define its spirituality as the aggregate of the spirituality of its members. This approach generalizes to nations, regions, and the entire world.

The spiritual state of a person is only known to himself as others do not know his state of internalization. To resolve this difficulty, one invokes the equivalence principle and the law of shadows. While the spiritual state of a person is not known, one can observe the shadow objects that correspond to his spirituality. Therefore, the interpersonal conduct of a person and his sharing practices are observed and their aggregate is taken over the whole community.

For the community to enjoy happiness, peace, and prosperity it is necessary that its level of aggregate spirituality be adequate. For that to happen, the spirituality for each member of the community needs to be adequate for at least a substantial majority within the community.

In old times the community used to expel or boycott its members who grossly fell off the expected moral behavior. Similarly, the church excommunicates its misbehaving members. The idea is to maintain a threshold of aggregate morality, though the practices are subject to misuse and overreach, because they violate the fundamental aspect that spirituality is not external morality, it is not externally imposed, rather it arises from within.

For a community to be spiritual three requirements are applied: the community needs to acknowledge spirituality as its goal at the community level; most of the community members must be adequately spiritual; and the community needs to have programs for the growth of spirituality of its members. Examples of such programs include the adoption of school curricula that support individual spiritual growth, state policies that promote genuine sharing, recognize truly altruistic contributions to the society, and encourage genuine sharing practices among its members.

As a corollary, it is inadequate just to look at economic indicators like the gross national product. It now becomes necessary to also look at spiritual indicators like the gross national happiness or gross national spirituality.

Lest a misunderstanding should arise, the above argument does not permit the state to select and promote a brand of values or behavior. That would be against the concept of spirituality, which is not externally induced and is an inward-looking view. We merely require the state to pay attention to spiritual indicators, and encourage its members to look inward.

The experience of oblivion demonstrates to man that he is part of other people and other people are part of him. Therefore, one's own happiness, fulfillment and peace derives from that of others. Similarly, peace and prosperity in one community derive from peace and prosperity in all other communities. A community cannot enjoy peace and prosperity in isolation, without other communities also enjoying the same.

The so-called problem of immigrants and their dehumanized treatment in present day is a reflection of spirituality missing from the communities, and adoption of spirituality is a solution to it. A spiritual community actively seeks worldwide peace and prosperity. The altruism quality for a community derives from the altruism quality of its members, and it will manifest in the community being an altruistically devoted citizen of the world.

One of the external manifestations of spirituality is peace. For the individuals it is peace within and peace with other people. It is similar with a spiritual community that enjoys peace within and peace with other communities. It can indeed be seen historically that nations have prospered while they maintained a spiritual emphasis in their lives, and they have withered when they lost this emphasis and became complaisant, greedy, and arrogant. World peace and prosperity lie in the adoption of a spiritual way within its communities.

Changes in Self

Man's own self represents the state of man at any given time. It is not static but changes as man discovers his own self and travels the spiritual spiral. These changes happen either in positive direction with increasing wisdom, or they happen in the negative direction under the corruptive greed forces.

Under the corruptive greed forces, man neglects to seek his own innate conscience. Instead, he greedily seeks more resources, partakes and promotes corruption, is oppressive and exploitative towards other people, promotes disruption, conflict and war, and destroys happiness, fulfillment and peace for his own self as well as others. When man follows this course, he destroys whatever spirituality he had, corrupts his worldview, is devoid of wisdom, and continues into the abyss of corruption and decadence. The only way out for such a person is to realize that he has no happiness and peace, take a pause from what he is doing, and have an honest conversation with his own self to perchance consult his conscience. I say perchance because generally consultation with conscience is unavailable to such a person because his conscience gets buried under the mounds of greed, corruption, and dehumanization. It generally takes long and sustained effort to unearth the conscience from underneath, by purifying his worldview. However, occasionally such a person can be jolted out by an unusual experience which may make the falsehood in his worldview evident for him to correct it. In such a circumstance the conscience may become available for consultation.

If the worldview of a person is reasonably clean, his own self may be wise enough to consult his conscience and to be guided with respect to what is good and what is bad. This guidance generally puts a man on the spiritual course whereby he seeks to know his own self, and resists the temptation for greed.

Just living life provides knowledge, for example, through experience with food, water, people, trees, earth, and sky. If the knowledge is internalized, it transforms knowledge into wisdom. The magic ingredients in wisdom include the passion for the peace and happiness of others.

The wiser a person becomes, more altruistic he becomes, and more engrossed he becomes in oblivion in which his own self begins to dissolve into everybody else's self. This is a progressive process and represents continuously evolving infinity of states of the self.

As the state of his oblivion progresses, the nature of things begins to transform, even transmute. The meaning of the terms inward and outward undergoes change, for what is the meaning of outward when your own self has dissolved into the self of those who were previously external? Meaning of happiness changes, because it has now transformed into the happiness of others. Similarly, the meaning of peace and fulfillment changes. What would be your own fulfillment after it has transformed into the fulfillment of those who persecute you?

As a person gets deeper and deeper into oblivion, the opposites tend to merge. For example, your fulfillment versus the fulfillment of those who persecute you, and your survival versus the survival of your enemies. Such apparent contradictions, and other such factors, change the meaning of what is wisdom? The meaning of spirituality is itself transformed. What is the meaning of getting to know your own self after your own self has dissolved into the self of those who love you, those who persecute you, and those who are your enemies? No logic works here, and no scientific knowledge helps here. That is because the meaning of everything has been transformed, and even transmuted. A person laughs at what he used to regard as happiness, fulfillment, and peace; and what he used to regard as enlightenment. The meaning of everything changes, but it does not change until you reach the appropriate spiritual state. And you do not know what spiritual states you will meet until you get close to them. They are not known until they manifest, until you actually witness them. Yes, at such stages no analytics work, only witnessing works.

This change of meanings does not happen once. It happens repeatedly and constantly, again and again. No meanings persist, and the old meanings can become meaningless after a person acquires new meanings. The new meanings are also a passing phase: the person just keeps approaching new states, new meanings, new realities, and new truths. It is bewildering when a person reaches there. Death will truncate the process so that only few will manage to reach there.

Contemplate on the journey of a seed and how many morphologies it assumes along its journey. During just one spiral of its cyclic existence, the seed goes through germination, a fledgling, a plant, a tree, a fruit, and a seed again. All the various existences of the seed are regarded as facts. The tree is in the seed and the seed is in the tree. All existences of man through the stages of his wisdom and the corresponding stages of his connectedness to the universe are facts; their realities and truths are more bewilderingly different and changeable than those that exist in one cycle of the journey of a seed.

There is a human perception of what one regards as real or true, based on one's worldview and the state of wisdom. The worldview is changeable as more and more life experiences accumulate. Accordingly, one's perception of reality and truth is also changeable. There is a global perception of sunrise and sunset; but does the sun rise, or does the sun set? At the time of Galileo Galilei, there was a perception of a stationary earth; but was the earth stationary? Many concepts about reality and truth have entered the human mind throughout the history. Not all concepts about reality and truth, throughout history, were valid. Many were, and still are, a result of external indoctrinations and dogmas. What people today regard as the reality and truth will change as these indoctrinations and dogmas change and people evolve in their worldview.

Religions have entertained such concepts as objective reality and ultimate truth. The philosophers have therefore discussed them. However, neither the religions nor the philosophers have realized the spuriousness of these concepts.

The journey through the spiritual world passes through a number of stages. Along this journey, man's perception of himself, of the physical world, of the spiritual world, and of the universe changes at each stage. For the practical purpose of this journey, all these perceptions are real and true for as long as they last. Once a particular stage of the journey is over, the traveler realizes that the reality and truth that he had perceived at that stage is also over. The next stage comes with its own discoveries, its own bewilderments, and its own perceptions. These stages follow each other endlessly because the pursuit is never-ending. Therefore, there is no such thing as reality or truth except in a phenomenological sense, which serves man well during a stage of his journey. There is no fixed reality nor is there an absolute truth; rather, both are perceptions based on the stage of wisdom. No two persons experience the journey into the spiritual world the same way; no two persons have the same perception of reality and truth.

The perception of reality and truth at an advanced stage of the journey is, however, ineffable. People often use poetic language and myth, which admit varying stages of understanding and interpretation that change with advancement in wisdom. The traveler experiences this as he moves farther and farther into oblivion. One can experience it in a witnessing sense; one cannot verbalize it, nor can one understand it or analyze it rationally. An attempt to understand it destroys it, because understanding implies analysis, and one cannot analyze a whole without breaking it apart.

In this context, "to understand" is different from "to know" or "to experience through witnessing". To understand is an analytical process; to know is an experiential process that witnesses what happens, without wanting to understand it. The how's and why's of a situation can, however, be demonstrated as part of a witnessing event.

One Existence

As the wisdom of a person approaches infinity, his state of oblivion progresses, and he realizes that nothing really exists by itself; all existence is part of a universal existence, which we rename as One Existence. His actions morph into a refined and advanced form of oblivion where he acts out of an inner compulsion, unconscious of the consequences. He acts out of an intrinsic connectedness of all existence.

In the limit of infinite wisdom, the progression of oblivion converges into a final state that I call One Existence. One Existence therefore derives like a limit of a progression as the wisdom tends to infinity. The journey of knowing one's own self *turns out to be the journey towards One Existence.*

I remark that within the span of a finite lifetime, attainment of infinite wisdom is only a slim possibility. Most people accomplished in spirituality may therefore not see One Existence clearly, though they may be able to anticipate it via imaginative extrapolation.

Please note that the above discourse is about man discovering One Existence during the journey of knowing one's own self. Man can discover One Existence only when his wisdom tends to infinity. However, One Existence can exist irrespective of whether or not man discovers it. The situation is not unlike the fact that the force of gravity exists irrespective of whether man has formulated a Theory of Gravity.

As wisdom increases oblivion dominates a person's conduct. First, he realizes that all people are one, next he sees that all living things are one, and finally he comes to the realization that all things are one, all things together are one unified thing, everything ultimately converges into One Existence. One Existence is where all sequences converge to. Looking this way, One Existence is the sum total of all the individual existences. The individual existences are varied including people, animals, plants, microbes, and inanimate things. Now think about other planets, stars, galaxies, parallel universes, and entities for which humans have no awareness or even imagination. All these existences come together in One Existence.

One Existence is therefore practically infinite. It is infinite because it includes an infinite number of entities; it is infinite because it includes an infinite number of species of entities; it is infinite because it includes an infinite number of locations; and it is infinite because the locations, species, and entities can include elements beyond current human awareness and even human imagination.

Now consider the summing and totaling process. What is this process? Is it simple counting; is it summing up infinite sequences; is it a mathematical integration; is it something that is not yet invented; or is it something of which humans currently have no awareness or even imagination?

When I talk about One Existence, all these infinities are included in it. Therefore, the nature of One Existence is beyond comprehension or even imagination.

However, a person can have experience of his own self interacting with One Existence, assuming there exists an interface through which humans can interact with One Existence, and One Existence can interact with humans. Such experience can reveal only the interface; it cannot reveal the nature of One Existence. In other words, the manifestations will be revealed without revealing the nature of One Existence.

Man discovers this interface as part of his spiritual seeking. Because spirituality is personal, we can infer that this interface is personalized as well, so that the experience of One Existence is personal; and as the advanced stages of oblivion are ineffable, so is the experience of One Existence. Any verbalization of the nature of One Existence is an attempt to describe what cannot be described. For many traditions it can be a breach of the spiritual code of conduct to verbalize such experiences.

Because One Existence derives as a limit at infinite wisdom when all people dissolve into One Existence, the attributes of man also are passed along to One Existence. I refer to these attributes by the same names as they are for man. However, for differentiation, I will capitalize them and add a prefix OE to the names of those attributes. Thus, One Existence has Attributes called OE-Existence, OE-Wisdom, and OE-Oblivion.

As there are infinities about One Existence, there are infinities also in the Attributes of One Existence. These infinities arise in various ways: the way the Attributes are defined and the way they can manifest as well as the scope of these manifestations.

I commented on the summation and totaling process for One Existence. Similar process needs to be applied also to the attributes of One Existence. This process needs to be applied to the existence, wisdom, and oblivion attributes of man to generate the OE-Existence, OE-Wisdom, and OE-Oblivion attributes of One Existence. For example, all individual wisdom attributes come together to produce the OE-Wisdom Attribute of One Existence. Further, there are also implicit attributes. For example, the wisdom attribute implicitly includes the knowledge attribute, the internalization attribute, and the sharing attribute. The same totaling process will apply also to the implicit attributes. For example, all individual knowledge attributes come together to produce the OE-Knowledge Attribute of One Existence. The attributes of One Existence are similar to that of man though they are infinitely more intense and meaningful in their capabilities, manifestations, and operational scope.

There can be helpful analogies to try to understand the Attributes of One Existence. For example, to understand the infinities of OE-Existence Attribute, one might imagine an infinite Lego structure representing the OE-Existence Attribute of One Existence. The wisdom attribute of man leads him to experience that he really is part of One Existence. Therefore, a man might experience himself as a piece of the Lego, consciously recognize his place in the scheme of the Lego, realize that he can fulfill himself only by recognizing and achieving his place with One Existence, and fit into the right place within the scheme of the Lego. In the end, there is nothing left of his own existence except its connectedness to One Existence. Self-discovery thus leads to connectedness to and self-dissolution into One Existence.

I caution that the above analogy is only for men of normal wisdom to envision an illustrative scenario. It has little to no meaning for those advanced into oblivion and actually witnessing the happenings. Further, such an analogy is necessarily of restricted helpfulness and should never be extended beyond.

One Existence follows if one assumes the existence of man and his internalization attribute which generates a bonding that leads to progressive oblivion experiences. In this sense, man is a passive creator of One Existence in the sense that One Existence would not be possible in the absence of man. Man does not actively create One Existence; rather, the general dynamics of wisdom and oblivion require that One Existence be present. The same dynamics also require that One Existence must have the attributes of OE-Existence, OE-Wisdom, and OE-Oblivion as they derive from the existence, wisdom, and oblivion attributes of man. According to these dynamics, *One Existence is created in man's image.*

Existence, wisdom and oblivion derive from the man's own self. One might ask how does man's own self arise? If one assumes that this is induced into man by One Existence, that would imply that One Existence creates man. We then get a paradoxical scenario in which *man creates One Existence and One Existence creates man.*

Therefore, it is not a fruitful pursuit to look for separate entities that play the roles of Creator and Creation. Both roles paradoxically exist in the same entity. Man is both a creator and a creation. One Existence is both a creator and a creation.

One Existence and man have shared attributes. Therefore, they work both ways. If their dynamics drive man to One Existence then similar dynamics will drive One Existence to man. Therefore, man is attracted towards One Existence and One Existence is attracted towards man. Given the infinitely more intense nature of the Attributes of One Existence, it can be inferred that the attraction of One Existence for man is infinitely more intense than the attraction of man for One Existence.

There is a story from Arabia that illustrates this relationship. An Arab was travelling on a camel, with all his livelihood necessities. He was passing through a desolate desert. He stopped and rested for a while. When he woke up his camel was gone, with all his livelihood. After a period of desperation, he saw his camel returning to him. There was no limit to the exaltation of the Arab traveler. It is stated that when a man fulfills his prerequisites and embarks upon spiritual seeking, One Existence is exalted far more than the Arab traveler was when he saw his camel coming back to him. The story is no exaggeration. After all, One Existence owes Its Existence to the existence of man. To express the same sentiment, it is stated that One Existence loves man far more than his own mother does.

In this symmetric relationship, man and One Existence play their roles without one of them placing obligations on the other. Moreover, the relationship is causal, determined by the dynamics of wisdom, even if wisdom appears to lose meaning in advanced stages of oblivion.

The limit of infinite wisdom may remain too far for most people, so that they may not come across One Existence. Therefore, they should proceed without One Existence: a spiritual stage cannot be witnessed before a person reaches there; and there is no reason to acknowledge it until he experiences it. There are, therefore, no hard requirements for his spiritual journey: every journey is different, every journey is special, and every journey is unique with or without One Existence.

Even when man does not witness One Existence, One Existence does witness man because the OE-Wisdom Attribute is all encompassing. Therefore, One Existence is still exalted more than the Arab traveler was upon seeing his camel return to him, when a man embarks upon seeking, even if he will never witness One Existence in his life: One Existence will still love him more than his own mother will, and do so unconditionally. After all, One Existence owes itself to the existence of man.

One Existence is an originality of my theory, it is not the old trodden God path.

Spiritual Sciences

The spiritual sciences investigate the laws that apply to the cause-and-effect relationships among entities and events in the spiritual world. This is similar to the physical sciences, which investigate the laws that apply to the cause-and-effect relationships among entities and events in the physical world. The equivalence principle puts the spiritual and physical sciences in direct correspondence with each other.

Three elements constitute the methodology of physical sciences, namely observation, analysis, and theorization. Since these elements are well-tested, I postulate that these three elements also apply to the spiritual sciences. The postulate is in accord with the equivalence principle.

There are two major difficulties in discussing the spiritual world. First is the complexity of the set of entities and events that populate the spiritual world. To describe these entities and events requires proportionately rich semantics and structures in the language. A similar difficulty was encountered in the physical sciences, and it was finally overcome by inventing the language of mathematics. The language of mathematics not only defined the terms unambiguously, it provided descriptions of the physical events with precision and clarity that one cannot imagine in natural languages.

Second difficulty arises from a diverse set of spiritual observations that people experience during their spiritual journey. Such observations can be short lived and unrepeatable. Therefore, the opportunity to express them using a structured language may be limited. One way to overcome this difficulty is to start the effort and to further develop the language framework along directions that the experience would seem to indicate. An example of such a start is the taxonomy of similarity classes of spiritual observations as was discussed in Axiom 2 on page 32.

Entities in the physical world include man, air, river, tree, earth, sun, planets, stars, and space. Entities in the spiritual world include man's own self, conscience, inner happiness, inner peace, and inner fulfillment, as well as entities opposite of these including tendencies for greed, corruption, oppression and exploitation.

Two very different types of entities populate the two worlds so that it means two very different things when the concepts of observation, analysis, and theorization are applied to a physical entity like air versus when they are applied to a spiritual entity like conscience.

Similarly, some events in physical world include sunrise, thundering, lightening, eclipse, and starburst. Some events in spiritual world include inner awakening, discovering the inner peace, becoming wise, and becoming altruistic.

These are two very different types of events so that the meanings of the concepts of observation, analysis, and theorization are very different in the two cases.

It is, therefore, necessary to find out the meaning of borrowing the elements of research methodology from physical sciences into spiritual sciences. The exercise is highly complex and the task is challenging. In conquering this complexity to meet the challenge I have spent substantial effort. I came to a realization that the three elements of research methodology that are borrowed from the physical sciences are inadequate to deal with the nature and scope of the entities and events in spiritual world. To construct even the very basic entities in spiritual world, such as self and conscience, I realized that additional elements of research methodology must be introduced. Therefore, I introduced two new elements of research methodology, namely the concepts of internalization (discussed under Axiom 4 on page 42) and witnessing (discussed under Axiom 5 on page 43). This expansion of the sciences is apparently adequate to deal with spiritual phenomena.

The need to expand the science methodology is an indication that the physical sciences are incomplete as they are.

Kurt Gödel's Incompleteness Theorem in mathematics proved in 1931 that limitations exist on the completeness and consistency of the physical sciences. Specifically, Gödel's incompleteness theorem makes a statement about mathematical integers, one of the simplest structures in mathematics, which can be paraphrased as below.

The properties of the integers cannot be consistently and completely described by a formal system containing a finite or infinite enumerable set of axioms.

In other words, such a formal system can describe the properties of integers such that the description can be complete, but then the description will not be consistent; or the description can be consistent, but then the description will not be complete. Thus, the description cannot be both complete and consistent. In other words, we can have a situation where a statement is known to be true without a proof that it is true, which represents incompleteness of the logical system. We can remove this incompleteness by postulating an axiom, namely that the statement is true. In that situation we would have to use an axiom that was not part of the original system of logic, which would imply inconsistency of the logical system. Thus, we can have either a system of logic that is incomplete but consistent, or a system that is complete but inconsistent.

If issues of incompleteness and inconsistency exist within such a rigorous science as pure mathematics, what claim can one lay about the completeness and consistency of the foundations of the rest of the physical sciences?

Further, there is Tarski's undefinability theorem, which he proved in 1933. In a simplistic way the theorem means that arithmetical truth cannot be defined within arithmetic. More generally, for any sufficiently strong formal system, truth in the standard model of the system cannot be defined within the system.

Tarski's theorem is not directly about mathematics but about the inherent limitations of any formal language sufficiently expressive to be of real interest.

Such limitations generally apply to physical sciences asserting that truth of a physical science cannot be expressed within the physical science. Truth is evasive unless the framework is extended.

Spiritual science extends the physical science in this sense.

Not many scientists know about Gödel's incompleteness theorem though it was proven long ago in 1931. Similarly, they are unaware about Tarski's undefinability Theorem. Scientist go about doing their science merrily unaware of the incompleteness and undefinability of the sciences. This unawareness of the scientists sometimes produces a level of arrogance of ignorance that does not stop them from making unwarranted claims about spirituality.

There are practical manifestations of this incompleteness and undefinability. Science presently stops at establishing theories and exploring their consequences. It does not extend to enriching life itself in terms of individual happiness, fulfillment and peace and worldwide peace and prosperity.

For example, Einstein discovered the famous equation $E = mc^2$ which forms the basis for the atomic bomb. The same Einstein had little qualms in writing to President F.D. Roosevelt urging him and guiding him to make an atomic bomb. The bomb was actually used twice on Japan with death and destruction of unimagined proportions. Copy of this letter is included below because of its historical importance and its consequences for mankind.

Such a disconnection between knowledge and actions too often exists among the physical scientists, because the physical science does nothing to guide life itself.

Therefore, I introduce Axiom 4 on page 42 that requires internalization of knowledge. As I have discussed earlier, internalization involuntarily stops a spiritual person from conducting himself in such a manner where a person's actions are disconnected from the responsibilities of his knowledge. A spiritual person works involuntarily to promote worldwide peace and prosperity.

Another illustration is occurring during the present time. The global climate catastrophe is taking place as I write this book. It too is happening because of the lack of internalization requirement in the physical sciences. Such a catastrophe would not be possible for a spiritual community.

Witnessing under Axiom 5 on page 43 is also an expansion of the physical sciences. It means a completely new way of acquiring knowledge and at the same time internalizing it without the use of observe-analyze-theorize cycle, and without using any rational processes. It is a vision that demonstrates phenomena as they take place in life. A crude example would be to know the taste of honey by actually tasting it versus inferring it from its descriptions. Witnessing provides a completely different view of phenomena in life. Instead of an analytic view, it affords a vision for a more unified and globally scoped view.

Albert Einstein
Old Grove Rd.
Nassau Point
Peconic, Long Island

August 2nd, 1939

F.D. Roosevelt,
President of the United States,
White House
Washington, D.C.

Sir:

Some recent work by E.Fermi and L. Szilard, which has been com-
municated to me in manuscript, leads me to expect that the element uranium may be turned into a new and important source of energy in the immediate future. Certain aspects of the situation which has arisen seem
to call for watchfulness and, if necessary, quick action on the part
of the Administration. I believe therefore that it is my duty to bring
to your attention the following facts and recommendations:

In the course of the last four months it has been made probable -
through the work of Joliot in France as well as Fermi and Szilard in
America - that it may become possible to set up a nuclear chain reaction
in a large mass of uranium,by which vast amounts of power and large quantities of new radium-like elements would be generated. Now it appears
almost certain that this could be achieved in the immediate future.

This new phenomenon would also lead to the construction of bombs,
and it is conceivable - though much less certain - that extremely powerful bombs of a new type may thus be constructed. A single bomb of this
type, carried by boat and exploded in a port, might very well destroy
the whole port together with some of the surrounding territory. However,
such bombs might very well prove to be too heavy for transportation by
air.

a64a01

-2-

The United States has only very poor ores of uranium in moderate quantities. There is some good ore in Canada and the former Czechoslovakia, while the most important source of uranium is Belgian Congo.

In view of this situation you may think it desirable to have some permanent contact maintained between the Administration and the group of physicists working on chain reactions in America. One possible way of achieving this might be for you to entrust with this task a person who has your confidence and who could perhaps serve in an inofficial capacity. His task might comprise the following:

a) to approach Government Departments, keep them informed of the further development, and put forward recommendations for Government action, giving particular attention to the problem of securing a supply of uranium ore for the United States;

b) to speed up the experimental work,which is at present being carried on within the limits of the budgets of University laboratories, by providing funds, if such funds be required, through his contacts with private persons who are willing to make contributions for this cause, and perhaps also by obtaining the co-operation of industrial laboratories which have the necessary equipment.

I understand that Germany has actually stopped the sale of uranium from the Czechoslovakian mines which she has taken over. That she should have taken such early action might perhaps be understood on the ground that the son of the German Under-Secretary of State, von Weizsäcker, is attached to the Kaiser-Wilhelm-Institut in Berlin where some of the American work on uranium is now being repeated.

Yours very truly,

A. Einstein

(Albert Einstein)

a64a02

Copy of August 2nd 1939 letter from Albert Einstein to President F. D. Rosevelt.

Closing Remarks

I offer some remarks in closing this treatise. The litmus test for an authentic spiritual living is the dynamic nature of happiness, fulfillment and peace that the spiritual traveler feels within his own self. Such happiness and peace evolve and grow. And as it grows over time, it witnesses ever new spiritual stages. The reality of the new stages overshadows the reality of the previous ones, and the new truths sometimes make one wonder how did he find the old truth satisfying. It is a garden of wonders and bewilderments.

This gradation comes with growth in wisdom. It does not happen if the wisdom does not grow. Even so, a spiritual life lived in consultation with one's conscience is far happier, fulfilling and full of peace than the life spent in pursuit of greed and false society mores. That is because the wisdom of a spiritual traveler does not remain static. The act of spiritual living itself makes the wisdom grow; at a minimum, in a relatively slow, even imperceptible ways.

Critical threshold in the spiritual living is something that comes naturally. Man's self is constantly seeking physical wellbeing and spiritual happiness. He is innately equipped to do that through his senses, intellect and conscience. It happens naturally and automatically. The monkey wrench comes in this smooth natural process through greed and false society mores. These are sicknesses, even if they appear invisible, like the termites that remain out of sight and yet eat into the foundation. This sickness is physical as well as spiritual. It keeps a person awake at night and perturbed during the day.

To save oneself from this sickness I have prescribed three simple steps: learn to trust your own self in a society where many are sick and blinded with greed; keep an open mind about the values that constitute your worldview; and keep searching and re-searching for the truth in values that form your worldview. This prescription will save you from the sickness and keep you physically healthy and spiritually at peace. Once you start on this path, the going is smooth and delightful, even ecstatic.

The three simple steps are enabled through scientific mindset. That is how foundational science is in spirituality. Though spirituality extends science and enriches it to help man live a happy, fulfilled and peaceful life, as it currently stands science is already a strong enabler for the spiritual way. Therefore, man should heed science and the mindset that it requires. With this heeding alone, man can live a happy, fulfilled and peaceful life. That is because science dispels falsehood and dogmas. The extensions to science, with the exception of internalization, become necessitated only towards the advanced spiritual stages. These correspond to the advanced stages of oblivion, and along this path the witnessing becomes important. However, man can go quite far along the spiritual spiral without feeling anything lacking in science plus internalization. The rest is the dynamics of the interpersonal spiritual conduct and the spiritual laws of wisdom.

These dynamics manifest along the long and winding road of traversing oblivion. This is where those spiritual manifestations become operative that leave science and scientific mindset behind. This is not because science becomes invalid, rather science becomes manifested in an entirely new mode and scope through witnessing. Starting the journey of witnessing is a major milestone along the infinitely winding and broadening spiritual spiral.

Very few people ever venture out in that domain. However, the present-day gurus of spirituality tend to spend most of their focus on descriptions of this domain that they themselves do not directly experience. Such focus is not genuine and it has little relevance to the journey of most spiritual travelers. In this treatise, I have dwelled on the path that is relevant to most spiritual seekers, and only cautiously pointed out any esoteric aspects that I have personally witnessed.